Herausgeber:

Prof. Dr. *A. Davison*	Department of Chemistry, Massachusetts Institute of Technology, Cambridge, MA 02139, USA
Prof. Dr. *M. J. S. Dewar*	Department of Chemistry, The University of Texas Austin, TX 7812, USA
Prof. Dr. *K. Hafner*	Institut für Organische Chemie der TH 6100 Darmstadt, Schloßgartenstraße 2
Prof. Dr. *E. Heilbronner*	Physikalisch-Chemisches Institut der Universität CH-4000 Basel, Klingelbergstraße 80
Prof. Dr. *U. Hofmann*	Institut für Anorganische Chemie der Universität 6900 Heidelberg 1, Tiergartenstraße
Prof. Dr. *K. Niedenzu*	University of Kentucky, College of Arts and Sciences Department of Chemistry, Lexington, KY 40506, USA
Prof. Dr. *Kl. Schäfer*	Institut für Physikalische Chemie der Universität 6900 Heidelberg 1, Tiergartenstraße
Prof. Dr. *G. Wittig*	Institut für Organische Chemie der Universität 6900 Heidelberg 1, Tiergartenstraße

Schriftleitung:

Dipl.-Chem. *F. Boschke*	Springer-Verlag, 6900 Heidelberg 1, Postfach 1780

Springer-Verlag	6900 Heidelberg 1 · Postfach 1780 Telefon (06221) 49101 · Telex 04-61723 1000 Berlin 33 · Heidelberger Platz 3 Telefon (0311) 822001 · Telex 01-83319
Springer-Verlag New York Inc.	New York, NY 10010 · 175, Fifth Avenue Telefon 673-2660

19 Fortschritte der chemischen Forschung
Topics in Current Chemistry

The Chemistry of Organophosphorus Compounds I

Springer-Verlag
Berlin Heidelberg GmbH 1971

ISBN 978-3-540-05458-0   ISBN 978-3-540-36549-5 (eBook)
DOI 10.1007/978-3-540-36549-5

Library of Congress Catalog Card Number 51-5497.

Contents

Synthesis of Organic Phosphorus Compounds from Elemental Phosphorus

Dr. L. Maier

Monsanto Research S.A., CH-8050 Zürich, Eggbühlstrasse 36

Contents

Introduction

Although the commercial manufacture of the element phosphorus did not develop until the 1830's and by 1844 the total phosphorus production in Great Britain was estimated at only 0.75 tons per year [73] already in 1845 Thenard [143] synthesized the first organophosphorus compounds by the interaction of calcium phosphide (or phosphorus and calcium) and methyl chloride. Today elemental phosphorus is readily and cheaply available. Its high reactivity makes it an excellent starting material for the synthesis of organic phosphorus compounds.

In spite of this only since 1950 was more closely looked at the direct synthesis of organic phosphorus compounds starting from elemental phosphorus. Rauhut, in his thorough review in 1963 on this subject [124], included 34 references. A year later, Grayson lectured on this subject [66]. Since that time the number of references concerning the synthesis of organic phosphorus compounds from phosphorus has more than quadrupled ($\sim$160) which attests to the lively interest in this field.

In the past few years methods have been found which allow the preparation of the following compounds starting directly from elemental phosphorus: primary, secondary, and tertiary phosphines, some of them having a very complicated structure, e.g., diphosphabarrellene (*10*),
aminophosphines,
stannylphosphines,
biphosphines,
cyclotetraphosphines,
phosphonium salts,
secondary and tertiary phosphine oxides,
phosphonous and phosphinous halides,
phosphonyl and phosphinyl halides,
phosphonic acids,
diphosphonic acids,
phosphinic acids,
secondary phosphites,
thiophosphites,
thiophosphonites,
thiophosphates,
phosphates, and
polyhydroxyphosphates.

It is the purpose of this review to summarize all the syntheses of organic phosphorus compounds which used elemental phosphorus as a starting material. The literature concerning the subject of this review is covered through January 1, 1971, including patent literature so far as abstracts are available in Chemical Abstracts.

I. Alkylation and Arylation of Phosphorus

1. Reaction with Alkyl and Aryl Halides

In 1861 A.W. Hofmann [72] isolated from the interaction of phosphorus *with ethyl iodide and zinc* in a sealed tube at 150 to 160 °C for several hours three products, e.g.

$[Et_3PH]I \cdot ZnI_2$,

$Et_3P=O \cdot ZnI_2$ and

Et_4PI.

Some years later, Carius [27] found that ethyl iodide also alkylates white or red phosphorus in the absence of zinc when the mixture is heated in a sealed tube to 150 – 170 °C. The following equation was proposed:

$$2 P + 4 EtI \longrightarrow PI_3 + Et_4PI$$

By a modification, i.e., opening the tube after the reaction, introducing ethanol, resealing and heating again to 160 °C, Et_3PO, 4 EtI and H_3PO_3 were obtained [27]. According to the above equation only 50% of the phosphorus should be converted to Et_3PO. Since, however, Crafts and Silva [33] obtained up to 73.5% Et_3PO (based on phosphorus reacted) and furthermore since Et_4PI can only be isolated after reduction with H_2S in 49% [103] the reaction probably corresponds to the following equation [103].

$$2 P + 7 EtI \longrightarrow Et_4PI_3 + Et_3PI_4$$

Interest in the preparation of tertiary phosphine oxides by this route has recently been revived by several groups of workers with the result being considerable improvement in the synthesis of tertiary phosphine oxides by this method (Kirsanov and coworkers) (for a summary see Table 1).

Thus heating red phosphorus and lower alkyl iodides (ratio 1.5:3) in the presence of a trace of iodine or diphosphorus tetraiodide in an autoclave at 200 °C for eight hours, followed by treatment with aqueous sodium sulfite or successive treatments with nitric acid and aqueous base gave Me_3PO (54%), Et_3PO (56%) and Bu_3PO (66%) [89].

Alkylations of phosphorus with higher alkyl iodides, [44,45,50] benzyl iodide, [89] and cyclohexyl iodide [46] did not have to be run in an autoclave. Hexyl iodide, for example, when refluxed with red phosphorus and small amounts of P_2I_4 (I_2 catalysis gave lower yields: 62%) for 30 hours gave, after treatment with sodium sulfite, tri-n-hexylphosphine oxide in 71% yield.

It has been suggested [45] that the reaction of red P and a trace iodine (or P_2I_4) with alkyl iodide at temperatures of up to 200 °C takes place not by a

Table 1. *Alkylation and arylation of phosphorus with hydrocarbon halides, thermally initiated*

Reactants	Temp. (°C)	Time (h)	Pressure	Products isolated	Ref.
$CH_3Cl + P_r + Cu$	360	10	–	CH_3PCl_2 (~16%); $(CH_3)_2PCl$, $(CH_3)_xPH_{3-x}$ addition of H_2 gives higher percentage $(CH_3)_2PCl$	61, 94)
$CH_3Cl + P_w + C/H_2$	300–400	2	+	$CH_3PCl_2 + (CH_3)_2PCl$ (68%)	78)
$CH_3Cl + P_w$	250	5	+	Me_4PCl (90%); CH_3PCl_2 (~10%); PCl_3	95)
$CH_3Br + P_r + Cu$	350	10	–	CH_3PBr_2 (20.2%); $(CH_3)_2PBr$ (8.4%)	94)
$CH_3Br + P_w$	230	8	+	Me_4PBr (90%); CH_3PBr_2 (~10%)	95)
$CH_3I + P_r + Cu$	280	17	–	CH_3PI_2 (13%)	94)
$CH_3I + P_w$	140	21	+	$Me_4PI + MeP_{14}I$	58)
$CH_3I + P_r + (I_2\text{-trace})$	200–220		+	$Me_3P{=}O$ (54%) a)	89)
$CHBr_3 + P_w$	190	1	+	$CHBr_2PBr_2 + (CHBr_2)_2PBr$	2)
$CF_3I + P_w$ (or P_r)	220–260	48	+	$(CF_3)_3P$ (84%); $(CF_3)_2PI$ (15%); CF_3PI_2 (1%) b)	13)
$CF_3CO_2Ag + P_r + I_2$	195	120	+	$(CF_3)_3P$ (50%); $(CF_3)_2PI$ (38%); CF_3PI_2 (12%)	25)
$CF_3I + P_r + Cu$	280	20	–	$(CF_3)_3P$ (2.5%); $(CF_3)_2PI$ (8.2%); CF_3PI_2 (3.4%)	94)
$CCl_3Br + P_w$	100	1	+	$CCl_3(Br)P\text{-}P(Br)CCl_3$ (41%) + $CCl_3(Br)P\text{-}PBr_2$ (43%)	19)
$CCl_4 + P_w$	157	104	+	CCl_3PCl_2(28%) + PCl_3 (10%) + P_r	117)
$ClCH_2OCH_3 + P_r + CuCl$	360	12	–	CH_3PCl_2 (3.7%) + $(CH_3)_2P(O)Cl$ (23.6%)	60)
$C_2H_5Cl + P_r + Cu$	440	23	–	$EtPCl_2 + Et_2PCl$	94)
$C_5H_5Br + P_r + Cu$	350	48	–	$EtPBr_2$ (4%) + Et_2PBr (trace)	94)
$C_2H_5Br + P_w$	210		+	$Et_4PBr + P_r \cdot x\ EtBr$	58)
$C_2H_5I + P_w$	150–170		+	$[Et_3PH]I \cdot ZnI_2$; $Et_3P{=}O \cdot ZnI_2$, Et_4PI, $Et_3P{=}O$ a)	27, 72)
$C_2H_5I + P_w$	175–180	24	+	Et_4PI, Et_3PI_2; isolated as $Et_3P{=}O$ (73.5%)	33)

Table 1 (continued)

Reactants	Temp. (°C)	Time (h)	Pressure	Products isolated	Ref.
$C_2H_5I + P_W$	180	24	+	Et_4PI_3, Et_3PI_4; isolated as Et_4PI (49%); $Et_3P{=}O$ [c)]	103)
$C_2H_5I + P_W$	200–220		+	$Et_3P{=}O$ (56%) [a)]	89)
$C_2F_5I + P_r$	219	40	+	$C_2F_5PI_2 + (C_2F_5)_2PI$	31)
$C_3H_7I + P_2I_4$	180–210		+	$PrPI_4, Pr_2PI_3 + Pr_3PI_2$	7)
$C_3F_7I + P_r$	200–220	8	+	$C_3F_7PI_2$ (30%) $+ (C_3F_7)_2PI$ (70%)	39)
$CH_2{=}CHCH_2I + P_r + I_2$	110–115	0.5	+	$(CH_2{=}CHCH_2)_3P{=}O$ (20–40%) [a)]	77)
$C_4H_9I + P_r + I_2$	200–220		+	$(C_4H_9)_3P{=}O$ (66%) also obtained from $I(CH_2)_4I$ [a)]	89, 50)
$I(CH_2)_4I + P_r + I_2$	190–200 then 205–210	1, 2	+ +	$\left[\underset{P}{\bigcirc\!\!\!\bigcirc} \right]^{+} I_3^{-}$ (40.5%) + polymeric products [a)]	36)
$i\text{-}C_5H_{11}I + P_r + I_2$	145	100	reflux	$(i\text{-}C_5H_{11})_3P{=}O$ (83.5%); $(i\text{-}C_5H_{11})_4PI$ (0.5%) [a)]	50)
$n\text{-}C_6H_{13}I + P_r$ or P_2I_4	205–210	5–9	reflux	$(C_6H_{13})_3P{=}O$ (82%) [a)]	44)
cyclo-$C_6H_{11}I + P_r + I_2$	215	2	+	$[(\text{cyclo-}C_6H_{11})_3PI_3]_2$ (95%)	46)
$n\text{-}C_7H_{15}I + P_r + I_2$ or P_2I_4	205–210	5–9	reflux	$(C_7H_{15})_3P{=}O$ (66%) [a)]	44)
$n\text{-}C_8H_{17}Br + P_W$	250–270	3.5	+	$n\text{-}C_8H_{17}PBr_2$ (28%) $+ (n\text{-}C_8H_{17})_2PBr$ (13%)	119)
$n\text{-}C_8H_{17}I + P_r + I_2$ or P_2I_4	205–210	5–9	reflux	$(n\text{-}C_8H_{17})_3P{=}O$ (78%) [a)]	44)
	200–210	10	reflux	$(n\text{-}C_8H_{17})_3P{=}O$ (90%) [a)]	45)
	180	50	reflux	$(n\text{-}C_8H_{17})_3P{=}O$ (90%) [a)]	45)
	150	100	–	$(n\text{-}C_8H_{17})_3P{=}O$ (5%) [a)]	45)
$n\text{-}C_8H_{17}I + I(CH_2)_4I + P_r + I_2$	210–230	2–2.5	reflux	$R_3P{=}O$ [a)] : Mono- and Poly-oxides (80%) isolated; $Bu_3P{=}O$ (8%); $(C_8H_{17})_3P{=}O$ (10%); $Bu_2(C_8H_{17})P{=}O$ (8%)	50)

Table 1 (continued)

Reactants	Temp. (°C)	Time (h)	Pressure	Products isolated		Ref.
$n\text{-}C_9H_{19}I + P_r + I_2$ or P_2I_4	205–210	5–9	reflux	$(n\text{-}C_9H_{19})_3P=O$ (57%)	a)	44)
$n\text{-}C_{10}H_{21}I + P_r + I_2$ or P_2I_4	205–210	5–9	reflux	$(n\text{-}C_{10}H_{21})_3P=O$ (77%)	a)	44)
$3,5,5\text{-}(CH_3)_3C_7H_{12}I + P_r + I_2$	205–210	1.5	reflux	$(3,5,5\text{-}(CH_3)_3C_7H_{12})_3P=O$ (70–75%)	a)	45)
$n\text{-}C_{16}H_{33}I + P_r + I_2$	205–210	1.5	reflux	$(n\text{-}C_{16}H_{33})_3P=O$ (70–75%)	a)	45)
$PhCH_2Cl + P_w$	300	4	+	$PhCH_2PCl_2$ (13.8%) + $(PhCH_2)_2PCl$ (0.7%)		119)
$PhCH_2Cl + P_r + Cu$	340–350		–	$PhCH_2PCl_2$ (7.0%)		59)
$PhCH_2Cl + P_w + CuCl$ (trace decalin)	170–180	11.5	reflux	$(PhCH_2)_4PCl$ (~77%); hydrolysis gave $(PhCH_2)_3P=O$ (55%) + $(PhCH_2)_2PO_2H$ (~5%)		28)
$PhCH_2Cl + P_w + CuCl +$ Dibutylcarbitol	188	50	reflux	$(PhCH_2)_2PO_2H$ (30–36%)	a)	28)
$PhCH_2Br + P_w$	150	1.5	–	$(PhCH_2)_4PBr$ (20%) + $PhCH_2PBr_2$ (30%)		144)
$PhCH_2I + P_r + I_2$	110–120	0.33	+	$(PhCH_2)_3P=O$ (81%) + $(PhCH_2)_2PO_2H$ (7%)	a)	89)
$p\text{-}ClC_6H_4CH_2I + P_r + I_2$	200–210		–	$(p\text{-}ClC_6H_4CH_2)_3P=O$ (~50%)	a)	89)
$p\text{-}CH_3C_6H_4CH_2Cl + P_w +$ CuCl (trace decalin)	170–180	11.5	–	$(4\text{-}CH_3C_6H_4CH_2)_3P=O$ (49%)	a)	28)
$3,4\text{-}Cl_2C_6H_3CH_2Cl +$ $P_w + CuCl$ (trace decalin)	170–180	11.5	–	$(3,4\text{-}Cl_2C_6H_3CH_2)_3P=O$ (24%)	a)	28)
$RC_6H_4CH_2Cl + P_r + I_2$ $R=Br, o,m,p\text{-}CH_3,$ $p\text{-}Et,Pr,i\text{-}Pr$	150	1	–	$(RC_6H_4CH_2)_3P=O;$ $R=p\text{-}Br$ (72%); $o\text{-}CH_3$ (73%); $m\text{-}CH_3$ (86%); $p\text{-}CH_3$ (82%); $p\text{-}C_2H_5$ (67%); $p\text{-}C_3H_7$ (73%); $p\text{-}i\text{-}C_3H_7$ (76%)	a)	159)
$C_6H_5Br + P_w$	(250)–315	(20)–4	+	$C_6H_5PBr_2$ (36%) + $(C_6H_5)_2PBr$ (31.9%)		94)

7

Table 1 (continued)

$C_6H_5Br + P_w$	300	4	$+$	$C_6H_5PBr_2$ (44%) + $(C_6H_5)_2PBr$ (22.8%)	119)
m-$CH_3C_6H_4Br + P_w$	300	4	$+$	m-$CH_3C_6H_4PBr_2$ (58%) + $(m$-$CH_3C_6H_4)_2PBr$ (20.8%)	119)
$C_6F_5I + P_w$	220	12	$+$	$(C_6F_5)_3P$ (high yield)	30)

a) After treatment of reaction mixture with Na_2SO_3 or NaOH or HNO_3; b) recycling three times and adding iodine gives $(CF_3)_3P$ (13%); $(CF_3)_2PI$ (19%); and CF_3PI_2 (18%); c) after reduction with H_2S.

radical mechanism but by an ionic mechanism. The reaction is not influenced by uv, or Bz_2O_2, and no radicals could be detected by esr. Furthermore, it has been said that the reaction proceeds by the initial formation of P_2I_4 which subsequently reacts readily with RI to give products of composition $[R_3PI_3]_2$ [46,52], for which structure 1 was proposed when R = cyclohexyl [46].

$$P + I_2 \longrightarrow P_2I_4 \xrightarrow{\text{(RI)}} R_6P_2I_6 + \text{(possibly } R_3PI_y) \xrightarrow{H_2O} R_3P=O$$

$$[(c\text{-}C_6H_{11})_3P \text{------} P(C_6H_{11}\text{-}c)_3]^+ I_5^-$$

$$\overset{\cdots}{I}$$

1

Indeed it is known for a long time, (Auger, 1904) that P_2I_4 reacts with alkylating agents [MeI, EtI, PrI] on heating to 180-210 °C to give after nitric acid treatment RPO_3H_2, R_2PO_2H ($^2/_3$ of the product) and $R_3P=O$ [7]. When dicyclohexyl-diidobiphosphine was treated with cyclohexyl iodide at 197-200 °C for 8 h, compound 1 was obtained in 33% yield. Benzyl iodide interacted with P_2I_4 already at 110 °C to give, after hydrolysis, tribenzylphosphine oxide in 92% and dibenzylphosphinic acid in 2% [89]. A similar reaction of $PhCH_2Cl$ with P_2I_4 at 130 °C for 17 h resulted in a 37% yield of the phosphine oxide [71]. Allyl iodide (20-40% $R_3P=O$) [77] and short [89] and long chain aliphatic alkyl iodides [44] gave with P_2I_4 tertiary phosphine oxides in yields of 53 to 75%. By reactions of red phosphorus, substituted benzyl chloride, and iodine at 120-130 °C several substitited tribenzylphosphine oxides have been obtained, after hydrolysis, in yields of 67-86% [159].

In a variation, high yields of phosphonium salts (77%) have been claimed to be obtained from the reaction of *white phosphorus and benzyl halides,* $XC_6H_4CH_2X$, with the aid of metal or metal salt (CuCl) catalysis [28]. The reaction may also be carried out in a high boiling solvent. Here dialkylation prevails and the phosphinic acid $(PhCH_2)_2P(O)OH$ is isolated after hydrolysis [28].

$$P_W + PhCH_2Cl + CuCl + \text{trace decalin} \xrightarrow[\text{(11.5 h)}]{\text{(170-180 °C)}} (PhCH_2)_4PCl \ (77\%)$$

No catalyst is necessary in the reaction of *methyl chloride or bromide with white phosphorus* at temperatures ranging from 220-310 °C and under pressure [95]. The best conditions for making tetramethylphosphonium chloride in over 90% yield is to heat a mixture of P_W : CH_3Cl = 1:1 to 1:2 to 250-260 °C for 5 h under pressure. Tetramethylphosphonium bromide is similarly obtained in 90% yield at 230 °C. The desired phosphonium salts are accompanied by methylphosphonous dihalide, phosphorus trihalide and sometimes a trace of dimethylphosphinous halide [95].

Since the latter products can be made the main products under different conditions [94] the following reaction scheme was suggested:

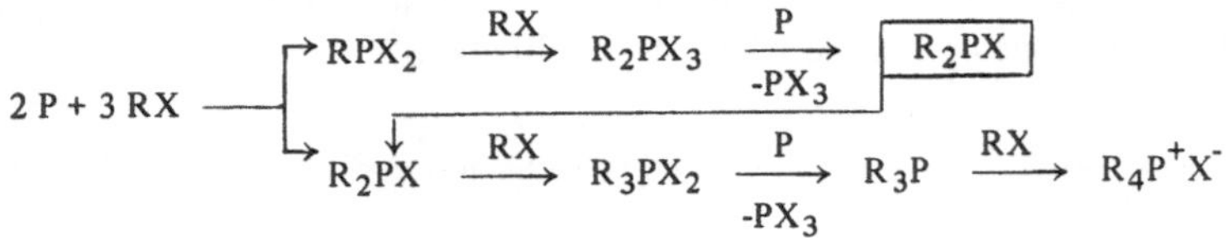

$$2\,P + 3\,RX \longrightarrow \begin{cases} \rightarrow RPX_2 \xrightarrow{RX} R_2PX_3 \xrightarrow[-PX_3]{P} \boxed{R_2PX} \\ \\ \rightarrow R_2PX \xrightarrow{RX} R_3PX_2 \xrightarrow[-PX_3]{P} R_3P \xrightarrow{RX} R_4P^+X^- \end{cases}$$

$$2\,P + 4\,RX \longrightarrow PX_3 + R_4P^+X^-$$

$$X = Cl,\ Br;\ R = CH_3$$

This scheme seems reasonable since it is known from the literature that alkyl and arylhalophosphines can be alkylated with alkyl iodides [7,37,71] or benzyl chloride [37,71] to give halophosphoranes and/ or phosphonium salts. Furthermore, reduction of halophosphoranes, R_2PCl_3 [142] and R_3PCl_2 [157] with elemental phosphorus to the threevalent state is also known.

The conversion of phenylene-bis (dichlorophosphine) (2) [112] and of alkyl [18,19] and aryldichlorophosphines [17] to phosphonium salts or, after hydro-

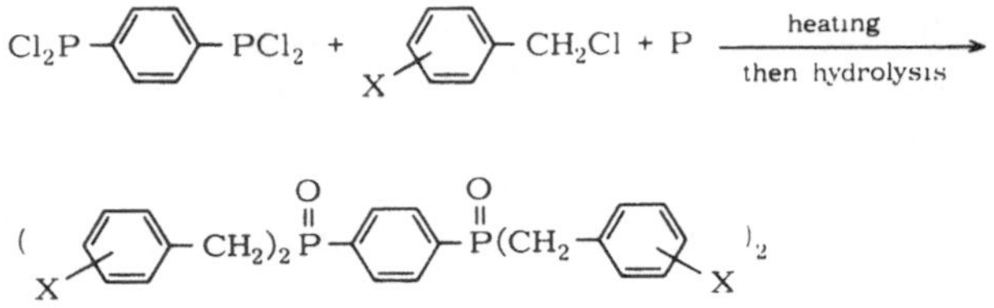

$$Cl_2P-\langle\ \rangle-PCl_2 \ + \ \underset{X}{\langle\ \rangle}-CH_2Cl \ + \ P \xrightarrow[\text{then hydrolysis}]{\text{heating}}$$

$$(\ \underset{X}{\langle\ \rangle}-CH_2)_2\overset{O}{\overset{\|}{P}}-\langle\ \rangle-\overset{O}{\overset{\|}{P}}(CH_2-\langle\ \rangle_X)_2$$

lysis, to phosphine oxides, by reaction with benzyl chlorides and elemental phosphorus seems to follow the same mechanism. The also formed PCl_3 [19] again indicates that elemental phosphorus acts as a reducing agent for halophosphoranes.

$$RPCl_2 + PhCH_2Cl + P \xrightarrow{(170\text{-}230\ ^oC)} PCl_3 + R(PhCH_2)_3PCl \xrightarrow{(OH^-)}$$

$$R(PhCH_2)_2P{=}O$$

same mechanism. The also formed PCl_3 [19] again indicates that elemental phosphorus acts as a reducing agent for halophosphoranes.

In a further variation, the primary products formed in the reaction of red phosphorus with alkyl iodides, may be *reduced with metals* such as Mg, Na, Li, Al, Zn, Fe, etc., first at 145-150 ^oC and then at 170 ^oC to give tertiary phosphines in 70-80% [51].

$$R_6P_2I_6 + 6\,Na \longrightarrow 6\,NaI + 2\,R_3P \xrightarrow{2\,R'I} 2\,[R_3R'P]I$$

1

If an alkyl halide is added to the reaction mixture after reduction but without isolating the trialkylphosphine, tetraalkylphosphonium halides are formed in 65 to 76% yield [51].

A cyclic phosphonium salt (*3*) was obtained, in addition to other products, when red phosphorus was treated with 1,4-diiodobutane at 210 °C for 3 h in the presence of catalytic amounts of iodine [36]. Alkylation of elemental phosphorus with hydrocarbon halides are summarized in Table 1.

A modification of the alkyl iodide reaction consists in the use of *alcohols and iodine* instead of alkyl halides as alkylating agents [43,47,48,49]. In a preliminary note it was reported [49] that the yields of tertiary phosphine oxides were, after hydrolysis, between 85-90%. Later it was, however, found that phosphoric acid monohydrate, which is a by product in this process,

$$ROH + P_{red} + I_2 \xrightarrow[\ 1.5\ h\ ]{\ 205\text{-}210\ ^oC\ } \xrightarrow{\ Na_2SO_3\ } R_3P{=}O$$

terminates the alkylation at an earlier stage, as a result of which alkylphosphonic acids are produced in 48 to 52% yield with higher alcohols $(> C_5)$ [43], whereas lower alcohols (C_3, C_4, cyclohexyl and phenylethyl) give phosphinic acids [47,48] in addition to tertiary phosphine oxides (see Table 2). An independent experiment with alkyl iodides, phosphorus and iodine in the presence of $H_3PO_4 \cdot H_2O$

$$ROH + P_r + I_2 \xrightarrow{\ reflux\ /\ NaOH\ } R_3PO + R_2PO_2H + RPO_3H$$

gave similar results [48].

While one group reported [45] that in the absence of iodine, red phosphorus and alkyl iodide (ratio 1:3) gave at 205-210 °C for 12 h only high molecular

Table 2. *Alkylation of red phosphorus with alcohols and iodine (ratio 1.2 : 1 : 3)*

Alcohols used	Temp. (°C)	Time (h)	Products isolated (%)	Ref.
n-C_3H_7OH	reflux	40–44	$Pr_3P{=}O$ (50.3) + $Pr_2P(O)OH$ (32–37)	47, 48)
n-C_4H_9OH	reflux	12	$Bu_3P{=}O$ (43) + $Bu_2P(O)OH$ (45)	47, 48)
n-$C_5H_{11}OH$	reflux	4–5	$Am_3P{=}O$ (43) + $AmP(O)(OH)_2$ (41)	48)
i-$C_5H_{11}OH$	reflux	4–5	$(i\text{-}Am)_3P{=}O$ (32) + $i\text{-}AmP(O)(OH)_2$ (50)	48)
n-$C_6H_{13}OH$	180°	30–32	$(n\text{-}C_6H_{13})_3P{=}O$ + $n\text{-}C_6H_{13}P(O)(OH)_2$ (48–52)	43)
c-$C_6H_{11}OH$	reflux	4	$(c\text{-}C_6H_{11})_3P{=}O$ (80)	48, 49)
n-$C_7H_{15}OH$	first: 100–110 then: 205–210	2–2.5 1–3	$(C_7H_{15})_3P{=}O$ (25–30) + $C_7H_{15}P(O)(OH)_2$ (48–52)	43, 49)
n-$C_8H_{17}OH$	first: 100–110 then: 205–210	2–2.5 1–3	$(C_8H_{17})_3P{=}O$ (25–30) + $C_8H_{17}P(O)(OH)_2$ (48–52)	43, 49)
n-$C_9H_{19}OH$	first: 100–110 then: 205–210	2–2.5 1–3	$(C_9H_{19})_3P{=}O$ (25–30) + $C_9H_{19}P(O)(OH)_2$ (48–52)	43, 49)
n-$C_{10}H_{21}OH$	first: 100–110 then: 205–210	2–2.5 1–3	$(C_{10}H_{21})_3P{=}O$ (25–30) + $C_{10}H_{21}P(O)(OH)_2$ (48–52)	43, 49)
3,5,5-$(CH_3)_3C_7H_{12}OH$	first: 100–110 then: 205–210	2–2.5 1–3	$(3,5,5\text{-}(CH_3)_3C_7H_{12})_3P{=}O$ (85–90)	49)
n-$C_{16}H_{33}OH$	first: 100–110 then: 205–210	2–2.5 1–3	$(C_{16}H_{33})_3P{=}O$ (25–30) + $C_{16}H_{33}P(O)(OH)_2$ (48–52)	43, 49)
$PhCH_2CH_2OH$	reflux	12	$(PhCH_2CH_2)_3P{=}O$ (79)	48)

weight resins, others [27,33,58,72,103] actually isolated phosphonium salts. Kirsanov and his coworkers [45] have suggested that tetraiodobiphosphine (either added or formed from $P_{red} + I_2$) is a required intermediate in the elemental phosphorus-alkyl iodide reactions catalyzed by iodine.

It is clear, however, that other evidence supports the direct attack mechanism and makes a radical chain mechanism likely, at least with alkyl chlorides and bromides.

The first characterization of simple *halophosphines* as the primary products of the direct reaction of white or red phosphorus with trifluoromethyl iodide at 200-220 °C under pressure was reported in 1953 by Bennett, Eméléus and Haszeldine [13].

$$P_4 + CF_3I \xrightarrow{200\text{-}220\ ^oC} CF_3PI_2 + (CF_3)_2PI + (CF_3)_3P + P_xI_y$$

The products can equilibrate under the reaction conditions; therefore the individual yields depend on the reactant ratios and the reaction time. Moreover, the individual product yields can be varied by the addition of one of the products [13]. Very likely, the electronegative fluoroalkyl groups prevent complete conversion to phosphonium salts by severely reducing the uncleophilic properties of the products. A radical mechanism was suggested.

Since CF_3I is prepared by the action of iodine on CF_3CO_2Ag, the *trifluoromethyl-substituted phosphines* may be prepared directly from silver trifluoromethylacetate, phosphorus, and iodine [25].

$$P + CF_3CO_2Ag + I_2 \longrightarrow CF_3PI_2 + (CF_3)_2PI + (CF_3)_3P$$

Extension of this reaction led to the preparation of perfluoroethyliodophosphines [31] and perfluoropropyliodophosphines [39] by raction of 1-iodo-perfluoroalkanes with red phosphorus under conditions indicated in the equations below.

$$P_r + C_2F_5I \xrightarrow[40\ h]{219\ ^oC} C_2F_5PI_2 + (C_2F_5)_2PI$$

$$P_r + C_3F_7I \xrightarrow[8\ h]{230\ ^oC} C_3F_7PI_2 + (C_3F_7)_2PI$$

$$\qquad\qquad\qquad\quad 30\% \qquad 70\% \qquad \text{(with 60\% conversion).}$$

While tris (perfluoropropyl) phosphine was not found even at 300 °C [39] the tertiary phosphine, tris (pentafluorophenyl) phosphine, and PI_3 were the sole products when a mixture of P_4 and C_6F_5I was kept at 220 °C for 12 to 14 h [30].

$$P_4 + C_6F_5I \xrightarrow[12-14\ h]{220\ ^oC} (C_6F_5)_3P + PI_3$$

It was shown by us [94] that the reaction of P with CF_3I is not limited to the condensed or liquid phase. Red phosphorus is also alkylated by *trifluoro-iodomethane vapor* over a copper catalyst [94]. The products are the same as in the liquid reaction but in reduced yield and different composition. The mono- and dialkylated compounds are the major products which is the reverse of the liquid system where the tris (trifluoromethyl)phosphine is usually the dominant product. This indicates that the phosphorus dihalide is a primary product of the reaction and not a decomposition or disproportionation product.

Furthermore, we could show [94] that the reaction is not limited to perfluoro-alkyl iodides. When methyl or ethyl halides were passed through a mixture of red phosphorus and copper at around 350 ^oC mainly phosphorus dihalides were produced. The phosphinous halide is formed in small amount along with some phosphines. With methyl bromide, e.g., the reaction product is composed of 80-90% CH_3PBr_2, 2-3% $(CH_3)_2PBr$ and 7-14% $(CH_3)_2PBr_3$.

$$P_{red} + RX \xrightarrow{(280-360\ ^oC)} RPX_2 + R_2PX$$

The phosphorane may have been formed either by addition of bromine – a pyrrolysis product of CH_3Br – to $(CH_3)_2PBr$ or more likely by addition of CH_3Br to CH_3PBr_2 (compare [37,71]). The overall yield

$$CH_3PBr_2 + CH_3Br \longrightarrow (CH_3)_2PBr_3$$

based on CH_3Br used is 28.6% [94,99], while that with CH_3Cl is only 16% [94]. A higher yield has been claimed to be obtained from the reaction of P_4–vapor, H_2 and CH_3Cl at 350 ^oC over an active carbon catalyst (68% yield) [78]. It was shown that the proportion of $(CH_3)_2PCl$ in the product can be increased from 4% to 35% by quicker removal of the products from the hot zone [94]. This indicates that $(CH_3)_2PCl$ is thermally unstable in the vapor phase which accounts for the different product distribution between liquid and vapor reactions. Regardless of the alkyl halide structure, catalyst or reaction conditions we seem to get the same initial products in all of the alkylation reactions.

The *reaction mechanism* of the copper catalyzed reactions is probably a radical chain process like that proposed by Rochow for the direct synthesis of organosilicon compounds [132].

$$RX + 2\ Cu \longrightarrow CuX + (RCu) \longrightarrow R\ ^\cdot$$

$$CuX + P_x \longrightarrow \cdot P_x X + Cu$$

$$XP\overset{.}{x} - \begin{array}{l} \overset{R\,\cdot}{\longrightarrow} \quad XP_x R \\ \boxed{RX} \\ \overset{}{\longrightarrow} \quad XP_x X + R \cdot \end{array}$$

Additional support for a free radical mechanism of the copper catalyzed reaction of red phosphorus with CH_3Cl or benzyl chloride comes from the isolation of methane, ethane, ethylene and propene in the CH_3Cl-reaction and of trans stilbene and toluene in the $PhCH_2Cl$-reaction, in addition to phosphonous dichlorides, $RPCl_2$ $(R = CH_3, C_6H_5CH_2)$ [59,61].

Curiously CH_3PCl_2 and $(CH_3)_2P(O)Cl$ were obtained from the interaction of red phosphorus and chloromethyl-methyl-ether in the presence of $CuCl$ at 360 $^{\circ}C$ [60].

$$ClCH_2OCH_3 \longrightarrow Cl\cdot + [\cdot CH_2OCH_3] \longrightarrow CH_2O + CH_3^{\cdot}$$

$$P_r + Cl\cdot + CH_3^{\cdot} \longrightarrow CH_3PCl_2 + (CH_3)_2PCl$$

$$(CH_3)_2PCl + CH_2O \longrightarrow (CH_3)_2P(O)Cl + [CH_2:]$$

The proposed reaction path seems unlikely. From the interaction of $(CH_2)_2PCl$ and CH_2O one would expect $(CH_3)_2P(O)CH_2Cl$ and not $(CH_3)_2P(O)Cl$.

Since reaction of phosphorus with alkyl halides occurs also without a catalyst, *direct attack by the alkyl halide* must be possible. Support that these reactions are also radical in nature has been provided by the work of Petrov, Smirnov, and Emel 'yanov [119]. In the reaction of white phosphorus with $C_6H_5CH_2Cl$ [119], $PhBr$ [94,119] m-$CH_3C_6H_4Br$ and n-$C_8H_{17}Br$ [119] in the liquid phase phosphonous dihalides are the predominant products but appreciable yields of phosphinous halides are also obtained. A correlation was observed between the tempera-

$$P_4 + 3\,C_6H_5CH_2^{\cdot} + 3\,Cl\cdot \longrightarrow C_6H_5CH_2PCl_2 + (C_6H_5CH_2)_2PCl$$

ture at which the reaction occurred and the stability of the radical [119].

$$PhCH_2\cdot < PhCHCH_3 < Ph_2CH\cdot < Ph_3C\cdot$$

$$300\ ^{\circ}C \qquad 270\ ^{\circ}C \qquad 250\ ^{\circ}C \quad 225\ ^{\circ}C$$

Benzyl bromide reacted with white phosphorus already at 150 $^{\circ}C$ to give benzylphosphonous dibromide (30%) and tetrabenzylphosphonium bromide (20%) [144]. (For other alkylation reactions see Table 1).

Phosphonous dihalides are nearly the exclusive products in the alkylation of white phosphorus with alkyl and aryl halides dissolved in a phosphorus trihalide,

15

which also takes part in the reaction [16]. The reaction is catalyzed by I_2, RI, Br_2, RBr and R_2PX (Bliznyuk, Baranov, Kabachnik).

$$P_4 + 2\,PX_3 + 6\,RX \longrightarrow 6\,RPX_2$$

The mechanism of the alkylation of white phosphorus in the presence of a phosphorus trihalide is probably the same as without PX_3. However, the phosphinous halides formed react with the phosphorus trihalide with the formation of phosphonus dihalides. The back reaction, i.e., formation of R_2PX and PX_3

$$R_2PX + PX_3 \rightleftharpoons 2\,RPX_2$$

from RPX_2 is well documented in the literature [12,137]. Therefore, this reaction seems to be reversible and the yield of phosphonous dihalide is raised by increasing of the amount of PX_3.

If we assume that initially not white phosphorus but PX_3 (which in view of the weak basic properties in less probable) is alkylated then formation of RPX_2 is expressed by the equation:

$$6\,RX + 6\,PX_3 \longrightarrow 6\,RPX_4 \xrightarrow{P_4} 6\,RPX_2 + 4\,PX_3$$

Since the basic properties increase with the number of alkyl substituents, phosphinous halides are more readily alkylated than P_w or PX_3. Catalysis by these compounds may therefore be expressed by:

$$RPX_2 + RX \longrightarrow R_2PX_3 \xrightarrow{P_4} R_2PX \xrightarrow{PCl_3} RPX_2$$

If no phosphorus trihalide is present the reaction should stop at the phosphinous halide stage or go even further to give tertiary phosphines. The preparation of phosphinous halides has been claimed [131]:

$$P_w + RCl + RPCl_2 \xrightarrow[4\ h]{340\ ^oC} RPCl_2 + R_2PCl$$

The reaction of white phosphorus with alkyl or aryl halides using PX_3 as *solvent* has recently been used to prepare *p*- $ClC_6H_4PCl_2$ (34%) [10], phenylene-bis (dichlorophosphine) (*3*) [10] (38%),

$$P_4 + 6\ p\text{-}ClC_6H_4Cl + 2\,PCl_3 + trace\ I_2 \xrightarrow[11\ h]{270-340\ ^oC} 6\ p\text{-}ClC_6H_4PCl_2$$

$$P_4 + 3\ p\text{-}ClC_6H_4Cl + 2\ PCl_3 + \text{trace } I_2 \xrightarrow[11\ h]{340\ ^oC} 3\ p\text{-}Cl_2PC_6H_4PCl_2$$

3

P-chloroisophosphindoline (*4*) [11] and P-chlorotetrahydroisophosphinoline (*5*) [9].

4

5

The synthesis of halophosphines using PX_3 as solvent are summarized in Table 3.

Several other *thermally initiated reactions* of white phosphorus with poly-haloalkanes have recently been reported.

Thus heating white phosphorus with CCl_4 at 157 °C for 104 h provides CCl_3PCl_2 in 28% along with PCl_3 (10%) and red polymeric solids (5%) [117].

$$P_W + CCl_4 \xrightarrow[104\ h]{157\ ^oC} Cl_3CPCl_2 + PCl_3 + P_{red}$$

While no higher alkylated product could be detected in the CCl_4 reaction [117], the formation of both, the phosphonous and phosphinous derivative was observed in the reaction of white phosphorus with bromoform at 190 °C for 1 h [2]. But also here the amount of the phosphonous dibromide was 4 to 6 times that

$$P_W + CHBr_3 \xrightarrow[1\ h]{190\ ^oC} CHBr_2PBr_2 + (CHBr_2)_2PBr + P_{red}$$

of the phosphinous bromide.

An unusual reaction is observed between white phosphorus and bromotri-chloromethane when they are heated to 100 °C for 1 h. The biphosphines *6* and *7* are obtained in 41 and 43% yield, respectively [1]. A summary of the thermally initiated alkylation and arylation reactions of phosphorus is given in Table 1.

Table 3. *Alkylation or arylation of white phosphorus with alkyl chlorides in PCl_3 or PBr_3 as solvent using I_2, Br_2, RI, or R_2PX as catalysts*

Reactants	Temp. (°C)	Time (h)	Products isolated (%)	Ref.
$C_4H_9Cl + P_w + PCl_3 + I_2$	290–320	5	$BuPCl_2$ (59.6)	16)
$n\text{-}C_6H_{13}Cl + P_w + PCl_3 + I_2$	290–310	5	$C_6H_{13}PCl_2$ (58.2)	16)
$n\text{-}C_9H_{19}Cl + P_w + PCl_3 + I_2$	290–320	5	$C_9H_{19}PCl_2$ (54.3)	16)
$PhCH_2Cl + P_w + PCl_3 + I_2$	230–240	10	$PhCH_2PCl_2$ (80)	16)
$PhCH_2Br + P_w + PBr_3 + I_2$	220–225	7	$PhCH_2PBr_2$ (20)	12)
$2,4\text{-}CH_3C_6H_4CH_2Cl + P_w + PCl_3 + I_2$	240–250	10	$2,4\text{-}CH_3C_6H_4CH_2PCl_2$ (70.3)	12)
$4\text{-}ClC_6H_4CH_2Cl + P_w + PCl_3 + I_2$	220–230	14	$4\text{-}ClC_6H_4CH_2PCl_2$ (68)	16)
$2,5\text{-}(CH_3)_2C_6H_3CH_2Cl + P_w + PCl_3 + I_2$	220–230	11	$2,5\text{-}(CH_3)_2C_6H_3CH_2PCl_2$ (62)	16)
$4\text{-}ClCH_2C_6H_4CH_2Cl + P_w + PCl_3 + I_2$	270–340	7–14	$4\text{-}Cl_2PCH_2C_6H_4CH_2PCl_2$ (75.3)	10)
$PhCH_2CH_2Cl + P_w + PCl_3 + I_2$	280–290	8	$PhCH_2CH_2PCl_2$ (47)	12)
(o-$C_6H_4(CH_2Cl)_2$ structure) $+ P_w + PCl_3 + I_2$	270–280		(bicyclic P–Cl structure)	11)
(o-$C_6H_4(CH_2CH_2Cl)$ structure) $+ P_w + PCl_3 + I_2$	230		(bicyclic P–Cl structure)	9)
$PhCl + P_w + PCl_3 + I_2$	320–340	7	$PhPCl_2$ (54.6)	16)
$PhBr + P_w + PBr_3 + I_2$	280–300	7	$PhPBr_2$ (71.0)	16)
$2\text{-}CH_3C_6H_4Br + P_w + PBr_3 + I_2$	230–250	8	$2\text{-}CH_3C_6H_4PBr_2$ (50.2)	16)

Table 3 (continued)

Reactants	Temp. (°C)	Time (h)	Products isolated (%)	Ref.
4-ClC$_6$H$_4$Cl + P$_W$ + PCl$_3$ + I$_2$	290–300	7–14	4-ClC$_6$H$_4$PCl$_2$ (34.5)	10)
4-ClC$_6$H$_4$Cl + P$_W$ + PCl$_3$ + I$_2$	340	11	4-Cl$_2$PC$_6$H$_4$PCl$_2$ (38.4)	10)
PhCl + P$_W$ + PhPCl$_2$	340	4	Ph$_2$PCl (50) + PhPCl$_2$ (12% additionally formed)	131)
PhCH$_2$Cl + P$_W$ + RPCl$_2$(R$_2$PCl) + I$_2$	170–230		R(PhCH$_2$)$_3$PCl [or R$_2$ (PhCH$_2$)$_2$PCl]	18)
XC$_6$H$_4$CH$_2$Cl + P$_W$ + ArPCl$_2$	170–200		Ar(XC$_6$H$_4$CH$_2$)$_2$P=O (after hydrolysis)	17)
PhCH$_2$Cl + P$_W$ + PhCH$_2$PCl$_2$ + I$_2$	170–220		(PhCH$_2$)$_3$P=O (after hydrolysis)	19)
XC$_6$H$_4$CH$_2$Cl + P$_W$ + 4-Cl$_2$PC$_6$H$_4$PCl$_2$	?		(XC$_6$H$_4$CH$_2$)$_2$P(O)C$_6$H$_4$(O)P(CH$_2$C$_6$H$_4$X)$_2$ (after hydrolysis)	112)
XC$_6$H$_4$CH$_2$Cl + P$_W$ + 4-Cl$_2$PC$_6$H$_4$PCl$_2$	170–220		[4-(XC$_6$H$_4$CH$_2$)$_3$PC$_6$H$_4$P(CH$_2$C$_6$H$_4$X)$_3$] 2Cl	20)

19

The reaction of white phosphorus with CCl_4 [117], $CHBr_3$ [2], CCl_3Br [1], and also with cyclohexane, cyclohexane-CCl_4 [5], and 1-chlorobutane [66] can also be initiated at lower temperature by visible light, or better *by radiation* from a cobalt-60 source (Henglein and coworkers). Normally the same products are formed as in the thermal reaction. For example, CCl_3PCl_2 is obtained on irradiation of a solution of P_4 in CCl_4 [117]. Since the yield is greater than 40 molecules/100 eV above 130 °C, a chain mechanism is involved. At lower temperatures a type of red phosphorus is produced as a consequence of free radical attack on white phosphorus. This red phosphorus contains a large number of radical groups from the solvent, e.g.,

$CCl_3P_7Cl_{0.3}$ (from P_w+CCl_4) [117],

$C_6H_{11}P_{4.5}$ (from P_w + C_6H_{12}) [5] and

$(C_4H_9)_4P_{39}Cl_2O_{10}$ (from P_w + C_4H_9Cl after oxidation) [66],

and may be used for chemical synthesis, e.g.:

$$CCl_3P_7Cl_{0.3} + Br_2 \longrightarrow CCl_3PBr_2 + PBr_3$$

It has been suggested that the initial step in the radiation chemical process involves the attack of a radical from the solvent on the P_4-molecule [1,2,5,66,117].

$$S \longrightarrow R \cdot + R' \cdot$$
$$R \cdot + P_4 \longrightarrow RP_4 \cdot$$

The radical RP_4 can either dimerize or propagate the chain by reacting with another solvent molecule:

$$2\,RP_4 \longrightarrow RP_8R$$
$$RP_4 + S \longrightarrow RP_4R + R' \cdot$$

Subsequent condensation of these intermediates leads to the formation of *"substituted" red phosphorus*. The solvents CCl_4 [117],

CHBr$_3$ [2] and C$_6$H$_{12}$/ CCl$_4$ [1] can, under the influence of ionizing radiation, react further with the corresponding "red" phosphorus to form low molecular weight products.

However, when CCl$_3$Br is used as solvent, no "red" phosphorus is formed because the CCl$_3$Br bond energy (49 Kcal/ Mol) is so low that the condensation reactions are dominated by competing degradation reactions [1].

The fact that the biphosphines 6 and 7 are formed and not the monomers CCl$_3$PBr$_2$ and (CCl$_3$)$_2$PBr indicates that they are either stable to CCl$_3$ radicals or, more likely, are reformed after radical attack according to the scheme [1]:

$$(CCl_3)(Br)\,P\text{-}P\,(Br)(CCl_3) + \cdot CCl_3 \longrightarrow CCl_3\text{-}\overset{|}{\underset{CCl_3}{P}}\text{-}Br + \cdot P(Br)CCl_3$$

6   8

$$8 + CCl_3Br \longrightarrow CCl_3PBr_2 + \cdot CCl_3$$

$$(CCl_3)_2PBr + CCl_3PBr_2 \longrightarrow 6 + CCl_3Br$$

In the radical initiated reaction of white phosphorus in the vapor phase with CF$_3$-radicals, generated from trifluoromethane and benzoyl peroxide, a cyclic four-membered phosphorus ring compound, tetrakis (trifluoromethyl) cyclotetra-phosphine (9) was formed [152]. Interaction of white phosphorus with fluoro-form and benzoyl peroxide in carbon disulfide solution also gave a small yield of the cyclotetraphosphine 9 [152].

$$4\ F_3C\cdot + P_4 \longrightarrow \begin{array}{c} F_3C\text{-}P \longrightarrow P\text{-}CF_3 \\ |\qquad | \\ F_3C\text{-}P \longrightarrow P\text{-}CF_3 \end{array}$$

9

Ultraviolet irridiation of solutions of white phosphorus in alkyl iodides and halobenzenes produced only organophosphorus polymers [82,83] containing phosphorus-carbon bonds. In some cases, oxidation of the polymers with nitric acid gave phosphonic acids in small amounts [83]. A summary of the radiation experiments is given in Table 4.

Table 4. Alkylation of phosphorus with alkyl halides initiated by peroxides, visible light or radiation with a Co-60-source

Reactants	Temp. (°C)	Time (h)	Initiator source	Products isolated (%)	Ref.
$CH_3I + P_W$	60–80	200–300	Hg-lamp	$CH_3PO_3H_2$ (0.4) [a]	82, 83)
$CHF_3 + P_W$	25		Peroxide	$(CF_3P)_4$	152)
$CHBr_3 + P_W$	25–140		Co-60-γ-rays	$CHBr_2PBr_2 + (CHBr_2)_2PBr + [CHBr_2P_3]_x$	2)
$CCl_3Br + P_W$	25		Co-60-γ-rays	$CCl_3(Br)P-P(Br)CCl_3(41) + CCl_3(Br)P-PBr_2(43)$	1)
$CCl_3Br + P_r$ [P_4Br from $P_4 + PBr_3$]	25		Co-60-γ-rays	$CCl_3(Br)P-P(Br)CCl_3$ (10) + $CCl_3(Br)P-PBr_2$ (90)	1)
$CCl_3Br + P_r$ [$(CCl_3P_6O)_n$]	25		Co-60-γ-rays	$CCl_3(Br)P-P(Br)CCl_3(\sim30) + CCl_3^i(Br)P-PBr_2(\sim35)$	1)
$CCl_3Br + P_r$ [$(CHBr_2P_3Br)_n$]	25		Co-60-γ-rays	$CHBr_2PBr_2 + CCl_3(Br)P-PBr_2 + CCl_3(CHBr_2)PBr$	1)
$CCl_4 + P_W$	130		Co-60-γ-rays	CCl_3PCl_2 (41) + PCl_3 (9.5)	117)
$C_2H_5I + P_W$	60–80	200–300	Hg-lamp	$EtPO_3H_2$ (4.1) [a]	82, 83)
$C_4H_9I + P_W$	60–80	200–300	Hg-lamp	$BuPO_3H_2$ (4.0) [a]	82, 83)
$C_4H_9Cl + P_W$			Co-60-γ-rays	$Bu_5P_{39}Cl_2$	66)
$i\text{-}AmI + P_W$	60–80	200–300	Hg-lamp	$i\text{-}AmPO_3H_2$ (4.3) [a]	82, 83)
$c\text{-}C_6H_{12} + P_W$	100		Co-60-γ-rays	$C_6H_{11}P_4/5 \xrightarrow{Cl_2} C_6H_{11}PCl_2$	5)
$c\text{-}C_6H_{12} + CCl_4 + P_W$	25–100		Co-60-γ-rays	$CCl_3PCl_2 + C_6H_{11}PCl_2 + P_x(C_6H_{11})y(CCl_3)_z$	5)
$PhI + P_W$	60–80	200–300	Hg-lamp	$p\text{-}NO_2C_6H_4PO_3H_2$ (0.32 based on P used) [a]	82, 83)

a) After oxidation of the reaction mixture with HNO_3 (the per cent yield given is based on the weight of polymer obtained).

2. Alkylation by Unsaturated Compounds

The preparation of substituted diphosphabicyclo-octatrienes or "diphosphabarrelenes" (10) was achieved by Krespan, McKusick and Cairns [86,88] by direct reaction of red phosphorus with fluorinated acetylenes in the presence of a catalytic amount of iodine at 200 °C for 8 h under pressure. The reaction might possibly go through a diphosphorin intermediate (9), since phosphorin readily

$$P_{red} + R_F C \equiv CR_F \xrightarrow[200°C]{I_2} \left[\begin{array}{c} R_F \quad P \quad R_F \\ \\ R_F \quad P \quad R_F \end{array} \right] \xrightarrow{R_F C \equiv CR_F}$$

9

10

$$R_F = CF_3 [88], \ ClCF_2CF_2 [86].$$

reacts with hexafluoro-2-butyne at 100 °C to give a substituted "phosphabarrelene" [104].

"Diphosphabarrelene" (10) was also obtained by reaction of red phosphorus with 2,3-diiodohexafluoro-2-butene at 210 °C [88].

Similarly, reaction of red phosphorus with tetrafluoroethylene in the presence of iodine at 220 °C produced octafluoro-1-iodo-phospholane (11, 4%) and octafluoro-1,4-diiodo-1,4-diphosphorinane (12, 27%) [87]. Free radicals have been suggested as intermediates in these reactions [87].

$$P_{red} + CF_2 = CF_2 \xrightarrow[220°C]{I_2}$$

11   12

Under severe conditions (300 °C, 1000 atm.) white phosphorus is said to react with an olefin and hydrogen to give small yields of tertiary phosphines [112]. The yields are substantially improved by the presence of alkyl iodides. Thus, the yields obtained with ethylene are indicated in the equation below:

$$P + CH_2{=}CH_2 + (CH_3I, \text{ trace}) + H_2 \xrightarrow[800\text{-}900 \text{ atm.}]{250 \text{ °C}/15 \text{ h}} Et_3P + Et_4PI$$

4%   17%

A patent claims that red phosphorus reacts with 1-pentene in the presence of $AlCl_3$ at 35 °C during 2 h to give a phosphorus containing olefin polymer suitable for use as a lubricating oil [69].

When phosphorus vapor with argon and ethylene, propylene, butene-1, propane, methane, ammonia or hydrazine was swept through a discharge tube, phosphine, PH_3, was produced as the major gaseous constituent in all cases [158]. Analysis of the methane-phosphorus discharge reaction indicated at least six products among which PH_3, CH_3PH_2 and ethane were detected [158], indicating a radical process:

$$P_{vapor} + CH_4 \xrightarrow{\text{discharge}} PH_3 + CH_3PH_2 + C_2H_6$$

Reaction of a lubricating oil with white phosphorus at 150-300 °C followed by oxidation with air gives a solution of oil soluble phosphorus compounds, believed to contain phosphonic acid groups [130]. A summary of the reaction of unsaturated compounds with phosphorus is given in Table 5.

3. Reaction with Alcohols, Phenols, and Amines

While red phosphorus (from $P_4 + PI_3$) gave only orange solids free of iodine when contacted with methanol or ethanol at room temperature for several days [115], a mixture of the corresponding alkylphosphines and phosphonium salts was produced when white phosphorus was heated with methanol or ethanol at about 250 °C for several hours [15] (Berthaud).

$$P_W + EtOH \xrightarrow{250\ °C} EtPH_2\ (\sim20\%) + Et_4P^+OH^-\ (20-30\%)$$

Phenol did not react with red phosphorus at 300 °C in the absence of water; but in the presence of small quantities of water, reaction at 250 °C gave small yields of phenylphosphine, diphenylphosphine, and phenylphosphonic acid [149]. Larger amounts of water favor the formation of $(PhO)_3P=O$ [149].

White phosphorus in contact with ethylamine in a sealed tube at room temperature turns first red, then dark-red, and finally gives a black precipitate of approximate composition $P_{5.81}H(C_2H_5NH_2)_{0.26}$ [141]. Other amines give apparently similar products [84]. Obviously the effect of amine consists mainly in the conversion of white phosphorus to another modification [84].

4. Reaction with Disulfides

White phosphorus is reported to react with dialkyl disulfides at 200 °C (2 h) to give the corresponding trithiophosphites, e.g., $(BuS)_3P$ (70% yield) [140]. This reaction may also be carried out in high boiling solvents such as hydrocarbons

Table 5. Alkylation of phosphorus by unsaturated compounds

Reactants	Temp. (°C)	Time (h)	Pressure (atm)	Products (%)	Ref.
CH_4 + P		in discharge tube		PH_3, CH_3PH_2, C_2H_6, P_{red}	158)
$CH_2=CH_2$ (or propylene, butene, propane) + P		in discharge tube		mainly PH_3	158)
$CH_2=CH_2$ + P_w + CH_3I [a] (in C_6H_{12})	250	4–15	800–980	Et_3P + Et_4PI	112)
$CF_2=CF_2$ + P_r + I_2 [a]	220	8	+	[cyclic structure] $P-I(4) + I-P-P-I$ (33)	87)
$CH_3CH=CH_2$ + P_w + CH_3I [a] (in C_6H_{12}/H_2)	250	15	1000	$PrPH_2$ + Pr_3P + Pr_3PHI	112)
$CH_2=CH-CH=CH_2$ + $P_w(Et_2O/H_2)$	300	15	1000	polymer of butadiene containing phosphorus	112)
$CF_3C{\equiv}CCF_3$ + P_r + I_2 [a]	200	8	+	[aromatic ring structure with CF_3 and P groups] (43)	88)
$CF_3Cl=ICCF_3$ + P_r	210		+		
$CH_3CH_2CH=CH_2$ + P_r + $AlCl_3$ [a]	35	2		olefinic polymer containing phosphorus	69)
C_6H_{10} + P_w + CH_3I/H_2 [a]	250	15	1000	$C_6H_{11}PH_2$, sec. phosphine $C_{13}H_{25}P$	112)

Table 5 (continued)

Reactants	Temp. (°C)	Time (h)	Pressure (atm)	Products (%)	Ref.
C_6H_{10} + P_W + $C_6H_{11}I/H_2$ [a]	250	15	1000	$(C_6H_{11})_4PI + [(C_6H_{11})_3PH]I$	112)
$ClCF_2CF_2C\equiv CCF_2CF_2Cl$ + P_r + I_2 [a]	220	12	+	$R_F = ClCF_2CF_2$	86)
Aromatic hydrocarbons having aliphatic side chaines + P_W	150–300			after oxidation: RPO_3H_2, oil soluble	130)

a) Acts as catalyst.

$$P_4 + 6\ RSSR \xrightarrow[\text{2 h}]{200\ ^\circ C} 4\ (RS)_3P$$

which do not posses mobile hydrogen atoms, e.g., kerosin, at a temperature of 170 to 210 $^\circ$C. The products have been claimed to be useful as lubricating oil additives [105].

The reaction of white phosphorus with dimethyldisulfide, initiated by *irradiation*, is composed of two steps: a) formation of red phosphorus with participation of the solvent; b) formation of $(CH_3S)_3P$ from red phosphorus. The radiation chemical yields are of the order of several 100 molecules/100 eV at room temperature which indicates a radical chain reaction [133].

$$CH_3SSCH_3 \xrightarrow{\gamma\text{-rays}} 2\ CH_3S \cdot$$

$$CH_3S \cdot + P_4 \longrightarrow CH_3SP_4 \cdot$$

$$CH_3SP_4 \cdot + CH_3SSCH_3 \longrightarrow CH_3S \cdot + CH_3SP_4SCH_3 \text{ etc.}$$

At low dose rate red phosphorus has the composition $(CH_3SP_2)_n$ [133]. While commercial red phosphorus does not react with Me_2S_2 with irradiation, red phosphorus, obtained from white phosphorus by irradiation in cyclohexane and having the composition $(C_6H_{11}P_{4.86})_n$ [5], reacts with Me_2S_2 when irradiated with ^{60}Co γ-rays to yield $(CH_3S)_3P$ and $C_6H_{11}P(SCH_3)_2$ in the ratio 3.6 : 1 [134].

$$C_6H_{11}P_{4.86} + CH_3SSCH_3 \xrightarrow{\gamma\text{-rays}} 3.6\ (CH_3S)_3P + C_6H_{11}P(SCH_3)_2$$

The yield of $\sim$ 1000 molecules/100 eV proves a radical chain reaction. Further radiation after suspended phosphorus has disappeared gives the thiophosphate $SP(SCH_3)_3$ and the thiophosphonate $C_6H_{11}P(S)(SCH_3)_2$ again by a radical chain process.

$$RP(SMe)_2 + CH_3S \cdot \longrightarrow RP(SCH_3)_3 \longrightarrow R(S)P(SMe)_2 +$$

$$[Me \cdot \xrightarrow{Me_2S_2} Me_2S + MeS \cdot]$$

The yield of $C_6H_{11}P(SCH_3)_2$ is drastically diminished if the red phosphorus is contacted with air before reaction with Me_2S_2. This indicates that the reactive sites in red phosphorus are those phosphorus atoms which are not only linked to phosphorus atoms but also to another group [134].

The ionic reaction of R_2S_2 with white phosphorus in a dipolar aprotic solvent (acetone, CH_3CN, DMSO) proceeds, unlike the thermal process, exceptionally smoothly under very mild conditions (25 $^\circ$C) in the presence of a base.

Yields of trithiophosphites are high ($\sim$90%) [154].

$$P_4 + 6\,RSSR \xrightarrow[\text{aprotic solvent}]{OH^-} 4\,(RS)_3P$$

Organic compositions claimed to be useful as insecticidal, rust-inhibiting, lubricant und fuel-oil additives were obtained on heating monoolefinic polymers, white phosphorus and S_2Cl_2 or SCl_2 to 195-200 °C for 4 h [107]. The reaction may also be initiated by ^{60}Co -γ -radiation at 15-90 °C [67].

II. Organic Phosphorus Compounds from Nucleophilic Attack on Phosphors

1. Base Catalyzed Reactions with RX, CH_2O, Olefins and R_2NCH_2OH

Reaction of white phosphorus with sodium hydroxide or sodium ethoxide in ethanol generates a dark red solution of an uncharacterized metastable phosphorus compound [106] which possesses nucleophilic properties [6]. The red product decomposes slowly even at 0 °C to hydrogen, PH_3 and sodium hypophosphite [106]. But addition of methyl iodide to the solution gives CH_3PH_2, $CH_3PO_3H_2$, $(CH_3)_2PO_2H$ and $(CH_3)_3PO$ [6]. Analogous products were reported from reactions of iso-amyl iodide.

$$P_4 + 6\,NaOH + 2\,RI \longrightarrow 2\,Na_2HPO_3 + 2\,NaI + 2\,RPH_2$$

Organic phosphorus compounds are also produced in the interaction of white phosphorus with an epoxide or an episulfide and an alcohol or mercaptane in the presence of alkaline catalysts at 25 °C to 200 °C [145]. In order to remove P-H bonds the reaction mixture is treated with formaldehyde and oxidized. The products are said to be useful as hardeners for epoxy resins or as antistatic agents and fire retardants.

High yields of bis (hydroxymethyl) phosphinic acid ($\sim 90\%$) have been claimed to be formed in the reaction of white phosphorus with formaldehyde in a basic alcohol/ water medium at 45-65 °C. Other aldehydes apparently react similarly [120].

Our own research [102] indicates, however, that the reaction product from CH_2O and P_4 consists actually of equal amounts ($\sim 30\%$ each) of hydroxymethyl-phosphonic acid (*13*), bis (hydroxymethyl)phosphinic acid (*14*), and methyl-hydroxymethyl-phosphinic acid (*15*). Small amounts of methyl-phosphonic acid and tris (hydroxymethyl)phosphine oxide were also detected [102].

$$P_4 + CH_2O \xrightarrow[OH^-]{CH_3OH/H_2O} \underset{13}{HOCH_2PO_3H_2} + \underset{14}{(HOCH_2)_2PO_2H} + \underset{15}{CH_3(HOCH_2)PO_2H}$$

Nucleophilic intermediates formed in the white phosphorus-hydroxide reaction are also trapped by electron deficient alkylating agents, such as acrylonitrile, acrylamide, ethyl acrylate [126,127], or vinylphosphonate [155]. Reaction of white phosphorus with KOH and acrylamide in aqueous ethanol resulted in the formation of tris (2-carbamoylethyl) phosphine oxide in 74%. An analogous reaction with acrylonitril gave tris (2-cyanoethyl) phosphine oxide (16) in 53% yield (based on P_4 used) [126], and with vinylphosphonate produced tris (diethoxyphosphonylethyl) phosphine oxide (17) in 100% yield [155]. Diethyl-, or

$$P_4 + 2\,KOH + 4\,H_2O + 9\,CH_2=CHCN \xrightarrow{CH_3CN} \underset{16}{3\,OP(CH_2CH_2CN)_3} + K_2HPO_3$$

$$P_4 + 2\,KOH + 4\,H_2O + 9\,(EtO)_2\overset{\overset{O}{\|}}{P}CH=CH_2 \xrightarrow{CH_3CN} 3\,OP[CH_2CH_2\overset{\overset{O}{\|}}{P}(OEt)_2]_3$$
$$\underset{17}{} + K_2HPO_3$$

diarylvinylphosphine oxides produced tris (oxophosphinoethyl) phosphine oxides when treated with P_4 and a base [156]. Reaction with ethyl acrylate gave the corresponding phosphine oxide only in 8% yield, along with 4% yield of the tertiary phosphine [126].

The red metastable product, produced from P_4 and KOH or KOEt, does not appear to be an intermediate in these tertiary phosphine oxide syntheses, since when separately prepared, gave only minor amounts of phosphine oxide. This result indicates that the olefinic compounds must attack an earlier intermediate. The initial step was thought to involve nucleophilic attack of hydroxide ion on tetrahedral white phosphorus to give a phosphide ion that subsequently underwent a Michael-Addition to the electrophilic unsaturated compounds present [126] (Rauhut, Bernheimer and Semsel).

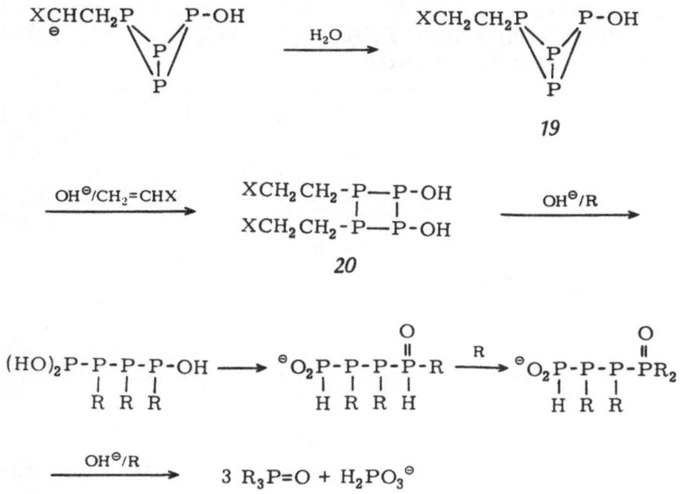

$$\begin{array}{c} \text{XCHCH}_2\text{P} \quad \text{P-OH} \\ \ominus \end{array} \xrightarrow{\text{H}_2\text{O}} \begin{array}{c} \text{XCH}_2\text{CH}_2\text{P} \quad \text{P-OH} \end{array}$$

19

$$\xrightarrow{\text{OH}^{\ominus}/\text{CH}_2=\text{CHX}} \begin{array}{c} \text{XCH}_2\text{CH}_2\text{-P—P-OH} \\ \text{XCH}_2\text{CH}_2\text{-P—P-OH} \end{array} \xrightarrow{\text{OH}^{\ominus}/\text{R}}$$

20

$$(\text{HO})_2\text{P-P-P-P-OH} \longrightarrow {}^{\ominus}\text{O}_2\text{P-P-P-}\overset{\overset{\text{O}}{\|}}{\text{P}}\text{-R} \xrightarrow{\text{R}} {}^{\ominus}\text{O}_2\text{P-P-P-}\overset{\overset{\text{O}}{\|}}{\text{P}}\text{R}_2$$

$$\xrightarrow{\text{OH}^{\ominus}/\text{R}} \quad 3 \text{ R}_3\text{P=O} + \text{H}_2\text{PO}_3^{\ominus}$$

Table 6. Dependence of the yield of tris(piperidinomethyl)phosphine oxide on the ratio of the reactants at 70 °C, using 15 g P

Ratio $P:CH_2O:HNC_5H_{10}$	Solvents	Reaction (h)	Yield of pure phosphine oxide in % (based on P_4 used)
1:2.25:2.25	60 ml H_2O + 150 ml EtOH	9	29.94
1:2.5:2.5	60 ml H_2O + 150 ml EtOH	9	37.5
1:3:3	60 ml H_2O + 150 ml EtOH	6.5	31.45
1:3.3:3.5	60 ml H_2O + 150 ml EtOH	9	26.6
1:2.25:2.25	60 ml H_2O + 150 ml CH_3CN	3	16.8
1:2.5:2.5	60 ml H_2O + 150 ml CH_3CN	2.75	32.6

31

Table 7. *Dependence of the yield of tris(piperidinomethyl)phosphine oxide on the solvent, using a ratio of $P_W:CH_2O:HNC_5H_{10}= 1:2.5:2.5$ at 70 °C (each time 15 g P were used)*

Solvent mixture	Reaction (h)	Yield of pure phosphine oxide in % (based on P_4 used)
60 ml H_2O	1.5 [a]	13.3
60 ml H_2O + 150 ml CH_3CN	2.75	32.6
60 ml H_2O + 150 ml C_2H_5OH	9	37.5
– 200 ml C_2H_5OH [b]	14	32.2
60 ml H_2O + 150 ml CH_3OH	9	19.4
60 ml H_2O + 200 ml i-C_3H_7OH	10	17.4
60 ml H_2O + 200 ml n-C_4H_9OH	10	21.4
60 ml H_2O + 150 ml THF	12	23.0 [c]
– 200 ml THF [b]	20	Reaction incomplete (P_W not completely used up)
60 ml H_2O + 70 ml Et_3N + 150 ml EtOH	3.5 [d]	24.2
60 ml H_2O + 150 ml C_6H_6	13	Reaction incomplete
60 ml H_2O + 150 ml $(CH_3)_2SO$	18	2.4
60 ml H_2O + 150 ml acetone	12	10

a) At 90 °C.
b) Paraformaldehyde.
c) In addition 3.7% tris(piperidinomethyl)phosphine were formed.
d) The pH decreases during the reaction from 9.7 to 8.1.

The absence of phosphonic and phosphinic acids seems to indicate that phosphorus is stable in its trivalent form in the intermediates *18, 19* and *20*. The instability of the pentavalent form was attributed to the increase in ring strain which would accompany the expansion of the bond angles in the transition from trivalent to pentavalent phosphorus [126].

No such restrictions are found in the reaction of white phosphorus with N-hydroxymethyldialkylamines [91,97,98]. Here, tertiary phosphine oxides comprise up to 45% of the reaction products; the rest is composed of phosphonic- and phosphinic acids and small amounts of secondary and tertiary phosphines [97] (Maier).

$$R_2NH + CH_2O \longrightarrow R_2NCH_2OH$$

$$R_2NCH_2OH + P_4 \xrightarrow{\ H_2O/EtOH\ } R_2NCH_2P(O)(OH)_2 + (R_2NCH_2)_2P(O)OH$$

$$+ (R_2NCH_2)_3P + (R_2NCH_2)_3P{=}O$$

The yields of the various products depend on the mole ratio of the reactants (Table 6), the solvents (Table 7), and in particular on the pH of the reaction mixture (Table 8) [97]. The highest yield of tris (piperidinomethyl) phosphine oxide, $(C_5H_{10}NCH_2)_3P=O$, is obtained in the uncontrolled reaction when the ratio $P : C_5H_{10}NCH_2OH$ is 0.5:1.25 and when 60 ml H_2O and 150 ml EtOH are used as the reaction medium. At a pH of ten or above the reaction occured very rapidly which makes initial alkoxide attack on phosphorus very likely [97]. This reaction, as suggested is reminiscent of the mechanism proposed for hydroxide attack on phosphorus [146]. The H-P<bond in (21) reacts very rapidly with R_2NCH_2OH to give $R_2NCH_2P<+ H_2O$ [101]. The phosphite bond is unstable and hydrolyzes to give a >P-OH bond [98].

$$R_2NCH_2OH + B \longrightarrow R_2NCH_2O^{\ominus} + BH^{\oplus}$$

21

Table 8. *Dependence of the phosphine oxide yield on the pH of the solution in the reaction P_W: $CH_2O:C_5H_{10}NH = 1:2.5:2.5$ (in H_2O - EtOH)*

pH		Temp. (°C)	Reaction (h)	$(C_5H_{10}NCH_2)_3P=O$ yield in %
11-7	a)	82	8	42.5
10	b)	82	1.5	4.1
7	b)	85	6	41.2
6	c)	85	11	11.2
4	c)	85	57	reaction incomplete

a) Uncontrolled reaction.
b) Maintained with NaOH.
c) Maintained with CH_3CO_2H.

$$>P\text{-}OCH_2NR_2 + H_2O \longrightarrow >P\text{-}OH + HOCH_2NR_2$$

Repetition of these steps leads to the final products. It is worth noting that the same products can be obtained independently therefrom, whether the intermediate with structure *20* rearranges to P^V or remains in the P^{III} state. It has namely been found that phosphorous and hypophosphorous acids also react with N-hydroxymethdialkalamines to give dialkylaminomethyl-substituted phosphonic and phosphinic acids [96], e.g.,

Table 9. *Organic phosphorus compounds from nucleophilic attack on phosphorus*

Reactants	Products (in %)	Ref.
$CH_3I + P_W + NaOH$ (NaOR)	$CH_3PH_2 + CH_3PO_3H_2 + (CH_3)_2PO_2H + (CH_3)_3P=O$	6)
$i\text{-}C_5H_{11}I + P_W + NaOH$ (NaOR)	$i\text{-}C_5H_{11}PO_3H_2$ (59%) + trace $(i\text{-}C_5H_{11})_2PO_2H$	6)
$CH_2O + P_W + NaOH$ (CH_3OH)	$(HOCH_2)_2PO_2H$ ($\sim$90)	120)
$CH_2O + P_W + NaOH$ (CH_3OH)	$HOCH_2PO_3H_2$ ($\sim$30) + $(HOCH_2)_2PO_2H$ ($\sim$30) + $CH_3(HOCH_2)PO_2H$ ($\sim$30) + $CH_3PO_3H_2$ (trace)	102)
Epoxide, Episulfide + P_W + base/alcohol	phosphorus containing products	145)
CH_2=CHCN + P_W + OH^-/CH_3CN	$(NCCH_2CH_2)_3P=O$ (53) + K_2HPO_3	126)
CH_2=CHCONH$_2$ + P_W + $OH^-/EtOH$	$(H_2NCOCH_2CH_2)_3P=O$ (74) + K_2HPO_3	126)
CH_2=CHCO$_2$Et + P_W + OH^-/CH_3CN	$(EtO_2CCH_2CH_2)_3P=O$ (8) + $(EtO_2CCH_2CH_2)_3P$ (4) + K_2HPO_3	126)
CF_2=CF$_2$, ClFC=CF$_2$, CH$_2$=CHSO$_2$CH$_3$ and CH_2=CHO$_2$CCH$_3$ + P_W + base	tertiary phosphine oxide + HPO_3^{2-}	127)
CH_2=CHP(O)(OEt)$_2$ + P_W + $OH^-/EtOH$	$[(EtO)_2(O)PCH_2CH_2]_3P=O(100)$ + HPO_3^{2-}	155)
CH_2=CHP(O)R$_2$ + P_W + $OH^-/EtOH$	$[R_2P(O)CH_2CH_2]_3P=O$, (R=Et, Bu, Ph) + HPO_3^{2-}	156)
$R_2NCH_2OH + P_W + H_2O/EtOH$ R=CH$_3$,Et,Bu,C$_4$H$_8$N,C$_5$H$_{10}$N C$_2$H$_4$OC$_2$H$_4$N, c-C$_6$H$_{11}$, 2,3, and 4-methylpiperidino	$(R_2NCH_2)_3P=O(\sim$40) + $(R_2NCH_2)_2PO_2H(\sim$20) + $R_2NCH_2PO_3H$ ($\sim$25) + $(R_2NCH_2)_3P$ ($\sim$5) + traces of $(HOCH_2)_2PO_2H$, $HOCH_2PO_3H_2$, and $R_2NCH_2PO_2H_2$	91, 97)
$RNHCH_2OH + P_W + H_2O/EtOH$ R=CH$_3$, Et, Pr	$RNHCH_2PO_3H_2 + (RNHCH_2)_2PO_2H$ + polymers	98)

34

$$2 R_2NH + 2 CH_2O + H_3PO_2 \longrightarrow (R_2NCH_2)_2P(O)OH + 2 H_2O$$

Since reaction of P_4 with R_2NCH_2OH also occurs at pH 7 or below, albeit much slower, a direct attack of R_2NCH_2OH on P_4 seems to be possible [97].

Table 9 summarizes the synthesis of organophosphorus compounds with base catalysis.

2. By Reactions with Organoalkali and Grignard Reagents

A new approach to the synthesis of dialkylamino-alkylphosphines consists in the reaction of white phosphorus with dialkylaminolithium compounds followed by treating the reaction mixture with alkyl halide; for example $(CH_3)_2NP(CH_3)_2$

$$2 P + 3 LiN(CH_3)_2 \longrightarrow LiP[N(CH_3)_2]_2 + Li_2PN(CH_3)_2 \xrightarrow[-3 \text{ LiCl}]{3 \text{ CH}_3\text{Cl}}$$

$$MeP(NMe_2)_2 + Me_2PNMe_2$$

and $[(CH_3)_2N]_2PCH_3$ were obtained in ~ 10 and 25% yield, respectively [93, 102]. Small amounts of trimethylphosphine and tris (dimethylamino)phosphine were detected by ^{31}P-NMR spectroscopy [102].

White phosphorus also reacts with carbon nucleophiles in ether or tetrahydrofuran to give dark red solutions believed to be complex organophosphides [128](Rauhut and Semsel). Hydrolysis of a mixture obtained from reactions of white phosphorus with organolithium and organomagnesium compounds gives the primary phosphine as the major product, with small amounts of secondary and tertiary phosphines being formed under some conditions.

$$C_6H_5Li + P_4 \longrightarrow [\text{organophosphide}] \longrightarrow C_6H_5PH_2 + \text{organopolyphosphine}$$

Under the best conditions found (PhLi:P_W = 1.6:1; Temp. 35-40 °C, Et_2O/THF) phenyllithium gave a 36% yield of phenylphosphine and phenylmagnesium bromide (PhMgBr : P_W = 1:1; Temp. 71 °C; THF) a 25% yield of phenylphosphine after hydrolysis. Use of a four fold excess of phenyllithium (PhLi:

P_w = 4:1, Temp. 25-30 ^oC, Et_2O) produced 15% diphenylphosphine in addition to 27% phenylphosphine. The alkylorganometallics, butyllithium and butylmagnesium bromide, produced in the reaction with phosphorus only 7.9% butylphosphine and 1% dibutylphosphine, while phenylsodium and triisobutylaluminum failed to produce simple compounds giving only polyphosphide products [128].

In all of these reactions the major by-product was an insoluble, non-melting, amorphous, yellow solid containing 40-60% phosphorus. The observation that the polyphosphide intermediates react further with additional phosphorus points to an analogy with the well known reaction of $(NH_4)_2S$ with sulfur to give ammonium polysulfides [128].

$$2 Li_2^+ [(C_6H_5)_2P_4]^{2-} + P_4 \longrightarrow 2 Li_2^+ [(C_6H_5)_2P_6]^{-2}$$

Like the red solutions obtained from P_w and NaOR, the red phosphides derived from phosphorus and organometallic reagents were found to react with alkylating agents to give tertiary phosphines [125]. Addition of butyl chloride to the reaction mixture from phenyllithium and P_w (ratio 2:2:1) in ether gave dibutylphenylphosphine (37%) and butyldiphenylphosphine (44%). Similar products were obtained from the reactions of 4-methoxyphenyllithium, 3-trifluoromethylphenyllithium or phenylsodium with phosphorus and butyl halides [125].

$$C_6H_5Li + P_4 + C_4H_9Cl \longrightarrow C_6H_5P(C_4H_9)_2 + (C_6H_5)_2PC_4H_9$$

1-Naphthyllithium and butyl halide, however, gave the secondary and tertiary phosphines after hydrolysis.

$$RLi + P_4 + BuX \longrightarrow RBuPH + R_2BuP$$

$$R = Bu \ 10\% \quad R = Bu \ 39\%$$

$$R = naphtyl \ 35\% \quad R = naphthyl \ 18\%$$

Although hydrolysis of the reaction mixture from phenylsodium and P_w produced no simple compounds, reaction with butyl chloride gave over 70% total yield of mixed tertiary phosphines. These results indicate the presence of structural units having two phenyl groups bonded to phosphorus. But the presence of alkali diphenylphosphides is unlikely since they would hydrolyze cleanly to diphenylphosphine.

Other electrophilic reagents also react with the polyphosphide ion [125]. Bis(2-hydroxypropyl)diphenylphosphonium bromide and 2-hydroxypropyldiphenylphosphine were isolated, both in 19% yield, by treating the reaction mixture from phosphorus and phenyllithium with propylene oxide. And reaction of the complex lithium phenylphosphide with benzaldehyde gave bis(α-hydroxybenzyl)

$$PhLi + P_4 \longrightarrow [polyphosphide] \xrightarrow{\overline{OCH_2\text{-}CHCH_3}} Ph_2PCH_2CHOHCH_3 +$$
$$[Ph_2P(CH_2CHOHCH_3)_2]^+Br^-$$

$$PhLi + P_4 \longrightarrow [polyphosphide] \xrightarrow[\text{2. } H_2O]{\text{1. PhCHO}} \overset{O}{\overset{\|}{PhP}}(CHOHPh)_2$$

phenylphosphine oxide in 12% yield [125].

In contrast to the reactions with the more reactive organoalkali compounds, the reaction of butylmagnesium bromide with phosphorus and butyl bromide (ratio 2:1:2) in refluxing tetrahydrofuran gave *tetrabutylcyclotetraphosphine* (22, R = Bu) in 42% yield [125,129] along with a 6% yield of Bu_2PH and a trace of Bu_3P. With a molar ratio of 6:1:10 the yield of $(BuP)_4$ dropped to 17%; in addition Bu_2PH (15%), Bu_3P (1%) and Bu_4PBr (10%) were also obtained in this case [125,129].

22

This simple, one step synthesis of cyclotetraphosphine seems to be general for aliphatic Grignard reagents. For example the ethyl and propyl homologs, $(EtP)_4$ and $(PrP)_4$, have been said to be formed in comparable yields by this method [32]. But reaction of iso-PrMgBr or cyclo-C_6H_{11}MgBr with white phosphorus and the appropriate alkyl bromide has been said to give solid compounds of composition R_2P_8 (R =iso-Pr, cyclo-C_6H_{11}) [3]. Commercial red phosphorus is reported not to react with Grignard reagents or with diethylzinc [115]. Freshly prepared red phosphorus (from P_w and PBr_3) containing Br-atoms as terminal groups, gave after treatment with CH_3MgI or Et_2Zn and oxidation with HNO_3 also only trace amounts of the corresponding acids [115]. Electrolysis of organomagnesium chlorides using a black phosphorus anode gave, however, tertiary phosphines [70]. A summary of the reactions of organometallic compounds with phosphorus is given in Table 10.

Table 10. *Reactions of organometallic compounds with phosphorus*

Reactants	Solvent	Temp. (°C)	Products (in %) (phosphines after hydrolysis)	Ref.
CH_3MgI, $PhMgBr$, $Et_2Zn + P_I$	Et_2O	35	no reaction	115)
BuLi or $BuMgBr + P_W$ (2:1)	Et_2O	25–35	$BuPH_2$ (7–9) + Bu_2PH (1)	128)
i-Bu_3Al, or $PhNa + P_W$	$C_6H_5CH_3$	45–100	only polyphosphides	128)
$PhLi + P_W$ (1.6:1)	Et_2O/THF	35–40	$PhPH_2$ (36)	128)
$PhLi + P_W$ (4:1)	Et_2O	25–30	$PhPH_2$ (27) + Ph_2PH (15)	128)
$PhMgBr + P_W$ (1:1)	THF	71	$PhPH_2$ (25)	128)
$(CH_3)_2NLi + P_W + CH_3Cl$	C_6H_6	70	$CH_3P[N(CH_3)_2]_2$ (25) + $(CH_3)_2PN(CH_3)_2$ (10)	102, 93)
$BuNa + P_W + BuCl$ [a] (2:1:2)	C_8H_{18}	35–40	Bu_2PH (13) + Bu_3P (1)	125)
$BuLi + P_W + BuBr$ [a] (2:1:1.5)	Et_2O	0–38	Bu_2PH (10) + Bu_3P (39)	125)
$PhLi + P_W + BuCl$ [a] (2:1:2)	Et_2O	40–42	$PhPBu_2$ (37) + Ph_2PBu (44)	125)
$PhLi + P_W + CH_3CH\text{-}CH_2$ [a] (2:1:2)	Et_2O	25–30	$Ph_2PCH_2CHOHCH_3$ (19) + $Ph_2P(CH_2CHOHCH_3)_2Br$ (19)	125)
$PhLi + P_W + PhCHO$ [a] (2:1:2)	Et_2O	25–35	$PhP(O)(CHOHPh)_2$ (12)	125)
$PhNa + P_W + BuCl$ [a] (2:1:2.4)	$C_6H_5CH_3$	45–50	$PhPBu_2$ (34) + Ph_2PBu (28)	125)
$4\text{-}MeOC_6H_4Li + P_W + BuBr$ [a] (2:1:2)	Et_2O	35–40	$4\text{-}MeOC_6H_4PBu_2$ (15) + $(4\text{-}MeOC_6H_4)_2PBu$ (20)	125)
$3\text{-}CF_3C_6H_4Li + P_W + BuBr$ [a] (2:1:2)	Et_2O	35–40	$3\text{-}CF_3C_6H_4PBu_2$ (36) + $(3\text{-}CF_3C_6H_4)_2PBu$ (37)	125)
$C_{10}H_7Li + P_W + BuBr$ [a] (2:1:2)	Et_2O	35–40	$C_{10}H_7PHBu$ (35) + $(C_{10}H_7)_2PBu$ (18)	125)
$EtMgBr + P_W + EtBr$	THF		$(EtP)_{4/5}$	32)
$PrMgBr + P_W + PrBr$	THF		$(PrP)_4$	32)
$BuMgBr + P_W + BuBr$ (2:1:2)	THF	71	$(BuP)_4$ (42) + Bu_2PH (6)	125,129)

Table 10 (continued)

Reactants	Solvent	Temp. (°C)	Products (in %) (phosphines after hydrolysis)	Ref.
BuMgBr + P_W + BuBr (6:1:10)	THF	71	$(BuP)_4$ (17) + Bu_2PH (15) + Bu_3P (1) + Bu_4PBr (10)	125)
i-PrMgBr + P_W + i-PrBr			$(i-Pr)_2P_8$	3)
c-C_6H_{11}MgBr + P_W + c-C_6H_{11}Br			$(c-C_6H_{11})_2P_8$	3)
Ph_4Sn + P_W (1:2)		235–250/ 12 h	$(Ph_2SnPPh)_3$ + $[(Ph_3Sn)_2P-P(SnPh_3)_2$ questionable]	136)
Ph_4Sn + P_W (1:2)		320/16 h	Ph_3P (85)	135)
$(C_6F_5)_2TlBr$ + P_W (3:2)		190/96 h	$(C_6F_5)_3P$ (70)	35)
Ph_3As + P_W (1:1)		300/4 h	Ph_3P [Ref. 139) reports no reaction]	81)
Et_3P, Ph_3P + P_W		60–70/ 100 days	$EtPO_3H_2$, $PhPO_3H_2$ (low yield after HNO_3 oxidation)	116)
$RhX(MR_3)_3$ + P_W	CH_2Cl_2	25	$RhX(MR_3)_2P_4$ (M=P, As)	90)

a) Alkyl halide added after completion of the reaction of P_W with RM.

3. By Reaction with Other Organometallic Compounds

Carbanions arising from other organometallic compounds also react with phosphorus with the formation of organophosphorus compounds. Heating tetraphenyltin with elemental phosphorus (1:2) for 16 h to 320 °C produced triphenylphosphine in 85% yield [135]. Proof for the intermediate formation of tin-phos-

$$\overset{\delta+}{Ph_3Sn} - \overset{\delta-}{C_6H_5} + P_4 \longrightarrow [Ph_3SnP \underset{\underset{P}{\overset{P}{|}}}{\diagup} PPh] \xrightarrow{2\ Ph_4Sn} 4\ Ph_3P + Sn_3P_n$$

phorus compounds was obtained when the reaction was carried out at 235-258 °C for 12 h. In this case it was possible to isolate $(Ph_2SnPPh)_3$, a yellow solid, [136]; (the also claimed formation of $(Ph_3Sn)_2P-P(SnPh_3)_2$ could not be confirmed [136a].

A 70% yield of tris (pentafluorophenyl)phosphine was obtained from the reaction of bromo-bis (pentafluorophenyl) thallium (III) with phosphorus at 190 °C for 4 days [35].

$$6\ (C_6F_5)_2TlBr + P_4 \longrightarrow 4\ (C_6F_3)_3P + 6\ TlBr$$

The report that triphenylphosphine is formed in quantitative yield when a mixture of Ph_3As and P is heated to 300 °C for 4 h [81] could not be confirmed by later workers [139].

Photopolymerization of withe phosphorus with UV-light (100 days at 60-70 °C) in the presence of tertiary phosphines (R_3P, R = Et, Ph) gave solid, insoluble polymers which contain organic radicals as terminal groups of the red P network. Oxidation of these polymers with HNO_3 gave small amounts of phosphonic acids [116]. It is concluded that commercial red phosphorus is a polymer with terminal groups composed of O and HO grouping. Thus red P, although containing as much as 99% P, is truly a compound, not an element in the true sense [116].

Recently an unusual reaction was reported. Low valent coordinatively unsaturated complexes of rhodium such as $RhX(MR_3)_3$, M = P, As, react with white phosphorus (S and Se) in dichloromethane solution to give complexes of the type $[RhX(MR_3)_2P_4]$. The complexes appear to contain an intact P_4 unit bonded to rhodium [90]. A summary of these reactions is included in Table 10.

4. By Reaction with Metals and Alkylating Agents

The first synthesis of organophosphorus compounds was achieved by the reaction of Ca_3P_2 with methyl chloride in 1845 by Thenard [143]. Ten years later Berlé [14] treated sodium phosphide with ethyl iodide at 100 °C for 6 h and obtained triethylphosphine and a product containing 67.2% iodine (possibly Et_3PI_2). Cahours and Hofmann [26] found it impossible to separate the products from the reaction of phosphorus with sodium and methyl iodide, but when zink instead of sodium was used and the mixture with phosphorus and ethyl iodide was heated in a sealed tube for several hrs to 150-160 °C, Hofmann [72] was able to isolate Et_4PI, $[Et_3PH]\,I\cdot ZnI_2$ and $Et_3P=O\cdot ZnI_2$.

In liquid ammonia, the alkali metals react with white phosphorus [41] and red phosphorus [21] to yield preferentially the biphosphides, M_2P-PM_2, which may be further reduced with NH_4Br to the monosodium phosphides, MPH_2, and characterized after alkylation with CH_3I as CH_3PH_2 [42], or which can be directly alkylated to give depending on reagent proportions used [21]: Me_2P –PMe_2 (17%), Et_2P–PEt_2 (28%), $EtPH_2$, Et_2PH (29%), $BuPH$ (34%) (the P-H compounds after an aqueous treatment), while subsequent treatment with sulfur gave $(Me_2PS)_2$ (26%), $(Et_2PS)_2$ (26%); $Me_2Et_2P_2S_2$ (20% meso-form), and $Et_2(PhCH_2)PS$ (20%).

In contrast to the liquid ammonia reaction trisodiumphosphide, Na_3P, is the major product resulting from the reaction of white phosphorus (red phosphorus proved less satisfactory) with sodium potassium alloy or with sodium dispersions in inert organic media (e.g. toluene) at temperatures varying from 80 to 145 °C [118]. The phosphide Na_3P reacts readily with methyl halides in glyme solvents to afford methylphosphorus compounds in ca. 60% overall yields under optimum conditions. A small amount (0,4%) of tetramethylbiphosphine was also isolated [118].

41

The claim that trialkylphosphine oxides are obtained by treating elementary phosphorus with sodium and alkyl halide during heating at 400 °C in an organic solvent is surprising [147].

From this type of reaction one would have expected the formation of tertiary phosphines.

5. By Electrochemical Methods

The electrolytic production, in all its variation, (different anode and cathode material different electrolytes, etc.) of phosphine is described in several patents [63]. It is believed that the reduction by this method proceeds via formation of the

$$P_W \xrightarrow[\text{(H}_2\text{O)}]{e^-} PH_3$$

tetraphosphine $[PH]_4$ [138].

As discussed recently in detail [100] PH_3 can readily be added to olefins and carbonyl containing compounds. Instead of carrying out this reaction in two different vessels, the process may be combined in one. One may then get not only primary, secondary and tertiary phosphines but also biphosphines and cyclotetraphosphines since reduction of P_W produces not only PH_3 but also P_4H_4, P_2H_4 and certain other highly reactive compounds [23,138] (Tomilov and coworkers). Thus electrolyzing white phosphorus in an aqueous buffered solution (CH_3CO_2Na), using a lead cathode, in the presence of styrene gives 2-phenalethylphosphine (23) (26.7%), tris (2-phenylethyl) phosphine (24) (1.7%), and the biphosphine tetrakis (2-phenylethyl) biphosphine (25) (4%).

$$P_W + PhCH=CH_2 \xrightarrow[\text{Pb-cathode}]{e^-/H_2O/CH_3CO_2Na} PhCH_2CH_2PH_2 + (PhCH_2CH_2)_3P +$$

$$\qquad\qquad\qquad\qquad\qquad\qquad\qquad\quad 23 \qquad\qquad\quad 24$$

$$(PhCH_2CH_2)_2P\text{-}P(CH_2CH_2Ph)_2$$

$$25$$

Furthermore, PH_3 ($\sim 20\%$) and an incompletely alkylated red, crystalline product is always obtained [75].

Running the electrolysis in THF gave 23 in 33.3% yield and 25 in 15.4%. Electrolysis in acetonitrile or methanol gives also small amounts of secondary phosphine ($PhCH_2CH_2$)$_2$PH, which is not formed in other solvents [75].

Electrolysis of organomagnesium chlorides in ethereal solution using a sacrificial black phosphorus anode, is reported to yield tertiary phosphines [70]. A tertiary phosphine oxide, tris (hydroxymethyl) phosphine oxide is produced in 64% yield (60% with respect to current) in the electrolysis of white phosphorus in a solution containing acetic acid, HCl, $(CH_3CO_2)_2Zn$, (to improve the conductivity) and formalin and using a lead cathode. Tris (α-hydroxyethyl) phosphine oxide was similarly obtained in 42% yield [114].

$$1/4\ P_4 \xrightarrow{+\ 3H^+,\ 3e^-} PH_3 \xrightarrow{3RCHO/H_2O} O{=}P(CHOHR)_3$$

It would seem that the first formed PH_3 reacts with CH_2O in acetic solution to give $(HOCH_2)_4PCl$ which is reduced electrolytically to the phosphine oxide.

Substituting cyclohexanone for formaldehyde in the white phosphorus electrolysis gives a secondary phosphine oxide in 21.3% yield (16.7% with respect to current) [113].

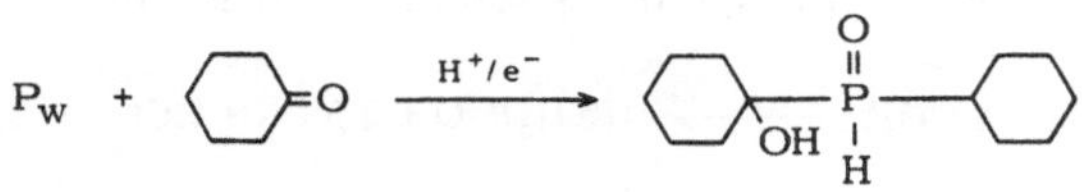

This result is not so surprising since it is known that PH_3 reacts with ketones in acetic solution also with the formation of phosphine oxides [24].

Electrolyzing white phosphorus in dimethylformamide solution in the presence of butyl bromide gave several organophosphorus compounds as shown below. Obviously dimethylformamide took part in the formation of methyltributyl-

$$P_W + BuBr \xrightarrow{(CH_3)_2NCHO/e^-} PH_3,\quad BuPH_2,\quad Bu_2PH,\quad Bu_3P{=}O$$

%	6.3	1.95	6.35	1.46

$$[Bu_2P]_2 + [BuP]_4,\quad CH_3Bu_3P_4,\quad Bu_3PO\cdot Bu_3PBr_2,\quad P_r$$

%	9.2	4.36	9.1	35.6

cyclotetraphosphine [74]. The total yield of organophosphorus compounds was about 28.5%.

Electrolysis of methanolic solutions of white phosphorus in the presence of alkyl halides and NaOH or NaOR gives a mixture of primary, secondary and tertiary phosphines (in 55% overall yield) when higher alkyls are used, and quaternary phosphonium salts (29-44%), when lower alkyls are the reactants [53].

$$P_W + C_9H_{11}I \xrightarrow{CH_3OH/CH_3ONa/e^-} C_9H_{11}PH_2 + (C_9H_{11})_2PH + (C_9H_{11})_3P$$

$$P_W + EtI \xrightarrow{CH_3OH/CH_3OK/e^-} Et_4PI$$

Electrolysis of a stirred suspension of red phosphorus in alcohol with graphite electrodes and with continous introduction of gaseous HCl gave trialkylphosphates and RCl. The following current yields were reported [148]:

$(RO)_3P=O$, R=CH_3 (48%, or 77.5% based on P),

Et (55, or 55),

Bu (58.7, or 88.3) and

Am (70, or 63.1).

An excess of HCl was essential as a deficiency of HCl caused a yield drop to some 14%.

Hydrogen was liberated at the cathode, while alkyl chlorides and $(RO)_3P=O$ were formed at the anode. The equation given for the reaction is [148]:

$$P_r + HCl + 4 ROH \longrightarrow (RO)_3P=O + 2.5 H_2 + RCl$$

It would seem that while this equation represents the overall process, the actual reaction taking place at the anode is the production of chlorine which reacts with phosphorus and alcohol to give the products observed (compare Ref. [56]). With allyl alcohol no consumption of phosphorus was observed and dichloropropanol was the sole product [148]. A summary of organophosphorus compounds produced by electrochemical methods is given in Table 11.

III. Organic Phosphorus Compounds from Electrophilic Attack on Phosphorus

White phosphorus is known to exist as a P_4 molecule [146] which is in a tetrahedral configuration containing an atom of phosphorus and an unshared pair of electrons at each apex. Therefore, this allotrope of phosphorus should be subject to easy attack by electrophilic reagents. It is somewhat surprising that only one such reaction has been reported [4]. When a solution of white phosphorus in carbon disulfide and one molar equivalent of $AlCl_3$ at -10 °C was treated with 1.5 molar equiv. of t-butyl chloride and the mixture hydrolyzed, a 30% yield of di-t-butylphosphinic chloride was isolated [4]. A second product was identified

Table 11. *Synthesis of organophosphorus compounds by an electrolytic procedure*

Reactants	Solvent	Products (in %)	Ref.
$RMgCl + P_{black}$ (anode)	Et_2O	R_3P	70)
$EtI + P_W$	$CH_3OH/NaOCH_3$	Et_4PI (29–44), isolated as Et_3PO	53)
$BuBr + P_W$	$(CH_3)_2NCHO$	$PH_3(6.3)$; $BuPH_2(1.95)$; $Bu_2PH(6.35)$; $Bu_3PO(1.46)$; $P_{red}(35.6)$; $Bu_3PO \cdot Bu_3PBr_2(9.1)$; $Bu_4P_2 + (BuP)_4$ (9.2); $CH_3Bu_3P_4$ (4.36)	74)
$C_9H_{11}I + P_W$	$CH_3OH/NaOCH_3$	$C_9H_{11}PH_2(27.6) + (C_9H_{11})_2PH(22.6) + (C_9H_{11})_3P(5.1)$ [isolated as acids and oxide]	53)
$ROH + P_r + HCl$	ROH	$(RO)_3P{=}O$; $R=CH_3(48)$; $Et(55)$; $Bu(58.7)$; $Am(70)$; $R=i\text{-}Pr$, $t\text{-}Bu$ (not isolated); $R=CH_2{=}CHCH_2$ gave $ClCH_2CHClCH_2OH$	148)
$CH_2O + P_W + acid$	H_2O	$(HOCH_2)_3P{=}O$ (64)	114)
$CH_3CHO + P_W + acid$	H_2O	$(CH_3CHOH)_3P{=}O$ (42)	114)
cyclo-hexanone $+ P_W$	$H_2O/AcOH/HCl$	$\underline{c}\text{-}C_6H_{11}(H)P(O)\text{—}\underset{HO}{\overset{\frown}{C}}H$	113)
$PhCH{=}CH_2 + P_W$	H_2O/CH_3CO_2Na	$PH_3(14)$, $PhCH_2CH_2PH_2(26.7)$, $(PhCH_2CH_2)_3P$ (1.7), $(PhCH_2CH_2)_4P_2$ (4)	75)
$PhCH{=}CH_2 + P_W$	$THF/H_2O/KOH$	$PH_3(17.7)$, $PhCH_2CH_2PH_2(33.3)$, $(PhCH_2CH_2)_4P_2(15.4)$, $(PhCH_2CH_2)_3P$ (5.1)	75)
$PhCH{=}CH_2 + P_W$	$CH_3OH/H_2O/KOH$	$PH_3(15.7)$, $PhCH_2CH_2PH_2(25.7)$, $(PhCH_2CH_2)_3P$ (9.1), $(PhCH_2CH_2)_4P_2(2.6)$, $(PhCH_2CH_2)_2PH$ (6.6)	75)
$PhCH{=}CH_2 + P_W$	$CH_3CN/H_2O/KOH$	$PH_3(14.0)$, $PhCH_2CH_2PH_2(14.3)$, $(PhCH_2CH_2)_2PH$ (7.0), $(PhCH_2CH_2)_3P(12.0)$, $(PhCH_2CH_2)_4P_2$ (3.4)	75)

$$P_W + t\text{-BuCl} \xrightarrow[\text{CS}_2]{\text{AlCl}_3} \xrightarrow{\text{H}_2\text{O}} (t\text{-}C_4H_9)_2P(O)Cl + (t\text{-}C_4H_9)_2P(O)H$$

as di-t-butylphosphine oxide. Other alkyl halides, RCl (R=t-amyl, i-propyl, n-Bu, cyclohexyl, n-octyl) and Lewis acids (FeCl$_3$, TiCl$_4$ but not ZnCl$_2$, ZrCl$_4$, or HgCl$_2$) also alkylate elemental phosphorus to give phosphinic acids or derivatives [4].

IV. Reaction of Phosphorus under Oxidizing Conditions

1. Reaction of Phosphorus with Olefins and Oxygen

It is known since a long time that phosphorus absorbs oxygen rapidly in the presence of olefinic compounds. The first observation of this type of reaction was probably made by Robert Boyle in 1681 [22] when he noticed that turpentine affected the oxidation of phosphorus. The glow accompanying the oxidation of phosphorus in air was not observed here. Graham [65] reported in 1829 that ethylene had a similar effect. Carbohydrates, oxalates and fatty acid salts are oxidized by air at 35 $^\circ$C when white phosphorus is present [29]. Several other reactions in which no identifyable product was isolated, have been summarized recently [62]. In reactions with ethylene [38,150], isoprene [150], cyclohexene [38,150,153], menthene [153], pinene [153], trimethylethylene [153], styrene [38,150], cholesterol [108], α-methylstyrene [38], octene-1, decene-1, dodecene-1, heptene-1 and hexadecene-1 [38,150], two atoms of oxygen are absorbed per atom of phosphorus and the products which separate during the reaction have a composition corresponding closely to the attachment of a P$_2$O$_4$ group to the double bond [150,153] or more correctly P$_2$O$_n$ ($4 \leqslant n \leqslant 5$) since n varies somewhat [38].

The oxygen content increases with increasing chain length of the olefin and approaches in reactions with 1-octene, 1-dodecene and 1-hexadecene the formula: olefin · P$_2$O$_5$ [38,150].

The reaction between white phosphorus, oxygen and cyclohexene in benzene solution yielded a product with the empirical formula C$_6$H$_{10}$P$_2$O$_4$ which hydrolyzed in the presence of oxygen to cyclohexene-1-phosphonic acid, (28) [150, 151,153]. Earlier reports [123,153] of an isolable intermediate product, olefin ·P$_2$O$_3$, could not be confirmed [38,150]. The yields of products are very high.

It was shown [150] that the reaction involves a free radical chain process, subject to catalysis by peroxides and α, α'-azobisisobutyronitrile, and inhibition by hydroquinone [150]. A kinetic chain length of at least 7000 at 50 $^\circ$C was estimated.

From potentiometric titration data it was deduced that the product $C_6H_{10} \cdot P_2O_4$ hydrolyzed readily to β-phosphitocyclohexanephosphonic acid (27)[151] which by β-elimination lost phosphorous acid (Walling and coworkers).

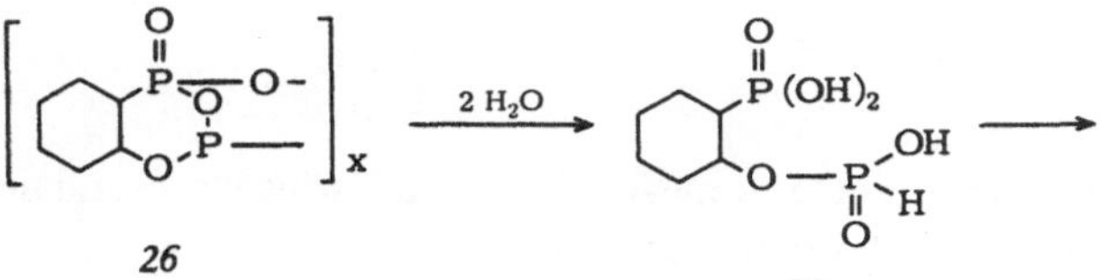

26   2 H$_2$O   27

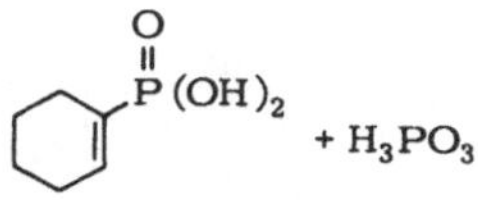

28

When isobutylene was used, $HOCH_2C(CH_3)_2PO_3 \cdot Pb$ was isolated suggesting an initial attack on olefin during oxidation involving a radical with partial structure P-O[151]. For the product from cyclohexene a polymeric anhydride structure containing one phosphorus per unit as a phosphonic anhydride and one phosphorus as a phosphite anhydride was proposed[151]. The formation of dialkylphosphites, $(RO)_2P(O)H$, by reaction of 26 with alcohols in 28-29% is in agreement with the proposed structure[151].

As further evidence for the polymeric nature of the reaction products the 1-hexadecene product $(C_{16}H_{32} \cdot P_2O_5)$[150] and the 1-octene product $(C_8H_{16} \cdot P_2O_5)$[38] gave molecular weights corresponding to five units.

Recent degradation studies indicate that in addition to 26 other structural units are present in the "phosphorate" product[38]. Thus, hydrolysis of the octene product $C_8H_{16} \cdot P_2O_5$ followed by esterification with diazomethane gave an octanediphosphonate with probable structure 29 and trimethylphosphate (up to 57% yield, based on P used). And treatment of the cyclohexene adduct $C_6H_{10} \cdot P_2O_4$, with PCl$_5$ followed by esterification allowed the isolation of the esters 30 to 33[38].

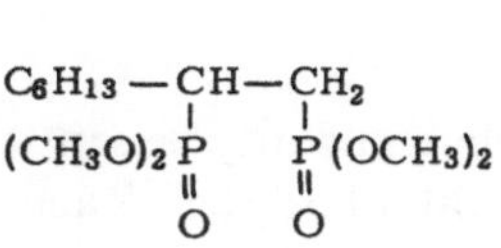

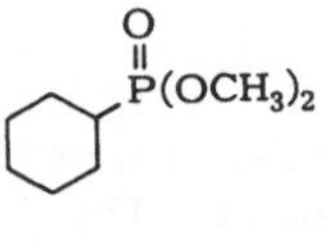

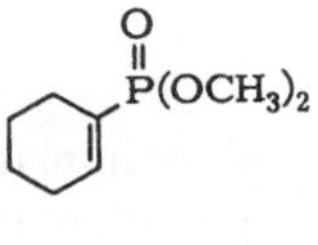

29   30   31

47

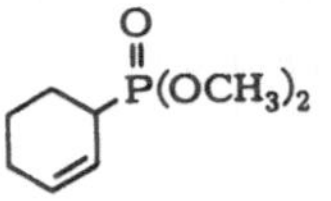

32

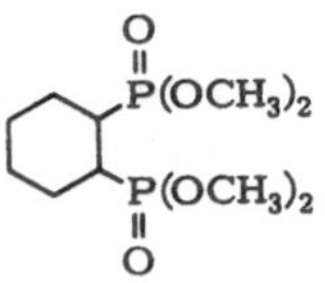

33

These results were interpreted to mean that the following structural units are present in the original "phosphorate" adduct:

34    35 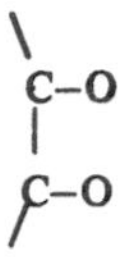   36    37

The isolation of diphosphonates points to the structural unit *34*, the unit *35* is contained in *26*. The formation of phosphates may indicate the presence of the unit *36*, but so far no diol was isolated. The isolation of the saturated cyclo-hexylphosphonate *30* confirms the presence of *37* [38] (Eckert, Hunger, Tavs).

The hydrolyzed products from the reaction of long chain olefins having even carbon numbers (C_8-C_{14}) with oxygen and phosphorus were said to have excellent surface active properties [110]. In contrast to Willstätter and Sonnenfeld [153] who reported that olefins used in excess of that required by the composition olefin ·P_2O_4 were not consumed, Cummins [34] found recently that excess olefin gives in addition to the "phosphorate" six- and eight-membered cyclic phosphonic acid esters, e.g. *38* and *39* when styrene is used. Unlike the "phosphorates"

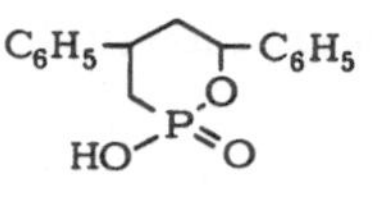

38

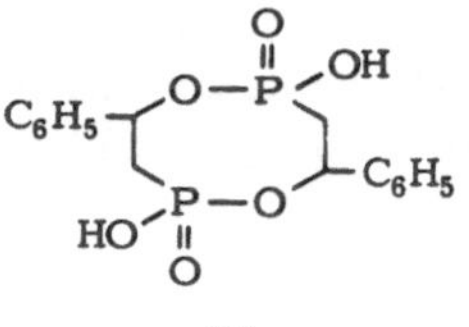

39

the cyclic phosphonates are stable to water and alcohols and may be separated from the "phosphorate" in this way. The absolute yields of *38* and *39* and the relative amounts of *38* and *39* in the crude product after removing "phosphorate" depend greatly on the styrene: phosphorus ratio (see Table 12). The percen-

Table 12. *Dependence of the ring size of cyclic phosphonates on the ratio of reactants and olefins in the reaction of white phosphorus with olefins*

Olefin	Ratio olefin: P_W	Yield %	Cyclic Phosphonates Distribution in %		"Phosphorate" olefin · P_2O_4 yield %
			6-mem. ring *(38)*	8-mem. ring *(39)*	
Styrene	0.5	22	24	76	74
	1.0	58	20	80	47
	2.0	75	48	52	26
	5.0	74	73	27	28
	7.7	70	83	17	31
Cyclohexene	7.7	61	64	36	24
α-Pinene	5.0	56	40	60	18
1-Octene	7.7	61	0	100	22

tage of *38* in the binary mixture of *38* and *39* increases from 24 to 83 as the styrene: phosphorus ratio is increased from 0.5 to 7.7 [34].

Like styrene, cyclohexene, α-pinene and 1-octene also formed the cyclic phosphonates, but in somewhat lower yields.

Distribution of 6- and 8-membered rings showed wide variation among the olefins examined (see Table 12). The esters were said to have surface-active properties and impart flame-retardant properties to cotton [34]. All the reported reactions of olefins with phosphorus and oxygen are summarized in Table 13.

2. Other Reactions of Phophorus under Oxidizing Conditions

When an olefin such as decene-1 is heated with white phosphorus and di-t-butyl peroxide the reaction is reported to follow the equation [57]:

$$x(\text{olefin}) + P_4 + (\text{t-BuO})_2 \xrightarrow[\text{7 h}]{150-160\ ^\circ C} R_xP_4O_2H_2 + 2\,C_4H_8$$

With 1-decene the value of x is between 4 and 6 under optimum conditions and the conversion of olefin and phosphorus is about 30%. The reaction is essentially complete in 3 hours. Small amounts of water accelerate the reaction.

The phosphorus product was not definitively identified. From a bromination experiment it was concluded that some double bond character remained in the hydrocarbon part. The product from 1-decene indicated about one acidic hydrogen for each two phosphorus atoms. Furthermore, the product appears to contain the tertiary phosphine structure, since treatment with CS_2 produced the red color characteristic of tertiary phosphines. Oxidation by nitric acid has been

Table 13. *Synthesis of organophosphorus compounds from white phosphorus, olefins and oxygen (or air) in aromatic hydrocarbons*

Olefin	Temp. (° C)	Time (h)	Products	Ref.
$CH_2=CH_2$	40	56	$C_2H_4 \cdot P_2O_4$	38,150)
$CH_2=CHCH_2OH$	20		$(C_3H_6O)_3 \cdot P_4O_6$	153)
$CH_2=CHCH_2OH$ (excess)	30	168	$(C_4H_9O_3P)$, cyclic phosphonate	34)
$ClCH=CHCl$	40		$C_2H_2 \cdot P_2O_4$ (containing Cl)	150)
Isoprene	40		Isoprene $\cdot P_2O_4$	150)
Methylmethacrylate	30	91	Cyclic phosphonates $+ (CH_2CMeCO_2Me)OPO(OH)]_x$	34)
Trimethylethylene			Trimethylethylene $\cdot P_2O_4$	135)
Cyclohexene	40	24	Cyclohexene $\cdot P_2O_4$	38,123,150,153)
Cyclohexene (excess)	40	93	Cyclic phosphonates	34)
Hexyne	40		$C_6H_{10} \cdot P_2O_4$	150)
1-Heptene	40	48	Heptene $\cdot P_2O_5$	38)
1-Octene	40	32–34	[Octene $\cdot P_2O_5$]5	38,150,110)
1-Octene (excess)	40	118	Cyclic phosphonates	34)
1-Decene	40	34	Decene $\cdot P_2O_5$	38)
1-Dodecene	40	34	Dodecene $\cdot P_2O_5$	38,150)
1-Dodecene (excess)	40		Cyclic phosphonates	34)
1-Hexadecene	40	40	[Hexadecene $\cdot P_2O_5$]5	38,150)
1-Hexadecene (excess)			Cyclic phosphonates	34)
9-Octadecene			Octadecene $\cdot P_2O_4$	110)

Table 13 (continued)

Olefin	Temp. (°C)	Time (h)	Products	Ref.
Vinyl-cetyl ether			Vinyl-cetyl ether · P_2O_4	110)
Polybutadiene			$C_{11}H_{18}O_4P$	34)
Styrene	40	32–34	Styrene · P_2O_4	38,150)
Styrene (excess)	30	160–165	Cyclic phosphonates	34)
α-Methylstyrene	40	91.5	α-Methylstyrene · P_2O_4	38)
Menthene			Menthene · P_2O_4	153)
α-Pinene			Pinene · P_2O_4	153)
α-Pinene (excess)	30	166	Cyclic phosphonates	34)
Cinamic acid ester	20		Cinamic acid ester · P_2O_4	153)
Oleic acid	20		Oleic acid · P_2O_4	123,153)
Limonene	20		Limonene · P_2O_4	153)
Poppy oil	20		Poppy oil · P_2O_4	153)
Colesterol	60		Colesterol · P_2O_4	108)
Olive oil	20		Olive oil · P_2O_4	153)

said to produce phosphonic and phosphinic acids and only 7% phosphoric acid indicating that the remainder is bonded to carbon. A sulfurization experiment indicated that the product also contained P-P bonds [57].

To account for the formation of isobutylene in as high as 92% yield with $(t\text{-BuO})_2$ a free radical initiation was proposed [57].

In the same manner as with olefin, white phosphorus also reacted with carbonyl compounds in the presence of oxygen to give products of apparently similar structure. For example with benzaldehyde it had the composition $(C_6H_5CHO \cdot P_2O_4)$ [111]. In the reaction with acetone, cyclohexanone, acetophenone and benzophenone, oxygen was absorbed at a slower rate, and the amount of oxygen absorbed and the yield were not quantitative. Hydrolysis of these products in Water gave phosphoric and another dibasic of unknown structure [111].

Addition of oxygen to a mixture of white phosphorus and an alcohol produces the corresponding dialkyl phosphite in 18-79% yields [40].

$$P_4 + ROH \xrightarrow{\;O_2\;} (RO)_2\overset{\displaystyle O}{\overset{\displaystyle \|}{P}}\text{-H}$$

$$R = Et\ (49\%);\ Pr\ (44\%);\ i\text{-}Pr\ (18\%)$$
$$n\text{-}C_4H_9\ (79\%);\ n\text{-}C_6H_{13}.$$

The reactions are either run in an excess of alcohol, or in benzene solution. They are exothermic and a reaction temperature of 50-75 $^\circ$C for 5 to 30 hrs is reported to be most satisfactory [40].

A recent patent claims that primary and secondary polyhydroxy phosphates are produced by admixing oxygen in at least the stiochiometric amount with white phosphorus and a polyhydroxy reactant such as a diol in the presence of a metal or metal oxide catalyst such as CuO, Al, Cu, Sc etc. [8]. The polyhydroxy

phosphates are claimed to be useful in the preparation of polyurethane foams [8].
The reactions are carried out at around 100 °C while passing air through the mixture for several hrs. Yields are between 70 and 80%.

Similar products are apparently obtained in the interaction of white phosphorus and oxygen with epoxides (epichlorohydrin, styrene oxide, and 1,4-butylene oxide) [64] or with phenols, polyhydric alcohols or mercaptanes using ethylacetate as solvent [76].

A simple way of preparing hydroxyethylidene-diphosphonic acid which is an excellent chelating agent for Mg and Ca-ions [92], consists in the reaction of white phosphorus with oxygen in acetic acid first at 80 °C and then heating the mixture to 150 °C [121].

$$P_W + CH_3COOH + O_2 \longrightarrow \underset{\underset{CH_3}{|}}{\overset{\overset{O}{\parallel}\;\;\overset{OH}{|}\;\;\overset{O}{\parallel}}{(HO)_2P-C-P(OH)_2}}$$

40

The reaction produces obviously a cyclic anhydride since a hydrolysis step is required to obtain the final product *40*. The yield is high approaching 70 to 80% [121].

Finally reaction of white (or red) phosphorus in excess alcohol (ratio 1:15) with chlorine yields trialkylphosphates [56].

$$P + 2.5\,Cl_2 + 4\,ROH \longrightarrow (RO)_3PO + HCl + RCl$$

The yield depends mainly upon the temperature used as shown in Table 14. The yellow flashing which is observed when chlorine is passed in, could be suppressed by diluting the chlorine with nitrogen. An excess of alcohol over the stoichiometric ratio 4:1 is believed to be necessary, its function being to complex the hydrogen chloride as the oxonium salt. Reduction of the ratio EtOH:P to 5.1 gave a chlorine containing product.

Table 14. *Dependence of the triethylphosphate yield in the reaction of P_W + ROH + Cl_2 on the temperature*

Reaction temp. (° C)	Yield of $(EtO)_3P=O$ (%)
-10 to 0	43
25–30	66
45–50	85
78 (reflux)	46

It has beem suggested that the mechanism is probably the same as that of the trialkyl phosphite reaction, i.e., chlorination of phosphorus to a chlorophosphonium species followed by alcoholysis and dealkylation. Whether PCl_3 or trialkalphosphites are actually present as discrete entities during the reaction is not

$$\geqslant P \xrightarrow{Cl^+} \geqslant P^+\text{-}Cl \xrightarrow{ROH} \geqslant P^+\text{-}OR \xrightarrow{Cl^-} \geqslant P=O + RCl$$

known [56]. The following phosphates were prepared: $(RO)_3 \overset{.}{P}=O$, $R=CH_3$ (61%), Et (85%), Bu ($\sim$ 100%), stearyl (98%) [56].

V. Use of Phosphorus as Reducing Agent in Organic Chemistry

It is outside the scope of this review to give a full account of the use of phosphorus as a reducing agent. However, a few reactions in which a phosphorus compound was produced as a by-product will be mentioned here. Red or white phosphorus reduces phosphorus (V) halides to the three valent state. This reaction has been used to prepare Ph_2PCl from Ph_2PCl_3 and red phosphorus [142], and Ph_3P from Ph_3PCl_2 and white phosphorus [95,157]. Other halogen containing compounds are similarly reduced, e.g., $PhSO_2NCl_2$ gives with red phosphorus $PhSO_2N=PCl_3$ [85], P_3NCl_{12} yields PCl_3 and $(PNCl_2)_{3/4}$ [54], RSCl produces RSSR [109], Cl_3CSH gives $CSCl_2$ [55], $PhCCl_3$ is converted to $PhCl_2C\text{-}CCl_2Ph$ [68] and derivatives of nitrobenzene are converted by phosphorus to the corresponding azoxy-compounds and amines [80], while nitrobenzaldehydes yield on treatment with red phosphorus the corresponding azoxybenzoic acids [79]. Reductions of other compounds with phosphorus have recently been summarized [79], p. 329 - 331) [108a].

VI. References

1) Airey, P.L.: Z. Naturforsch. *24b*, 1393 (1969)
2) –, Drawe, P.H., Henglein, A.: Z. Naturforsch. *23b*, 916 (1968)
3) Ang, H.G.: (cited by B.O. West) Record Chem. Progr. *30*, 249 (1969), ref. 10
4) Angstadt, H.P.: J. Am. Chem. Soc. *86*, 5040 (1964); Neth. Appl. 6 501 808 (1965); C.A. *64*, 6693b (1966)
5) Asmus, K.D., Henglein, A., Meissner, G., Perner, D.: Z. Naturforsch. *19b*, 549 (1964)
6) Auger, V.: Compt. Rend. *139*, 639 (1904)
7) – Compt. Rend. *139*, 671 (1904)
8) Baranauckas, C.F., Hodan, J.J.: U.S. Pat. 3 445 547 (1969)
9) Baranov, Yu.I., Fillipov, O.F., Varshavskii, S.L., Glebychev, B.S., Kabachnik, M.I., Bliznyuk, N.K.: U.S.S.R. Pat. 213 857; C.A. *69*, 77 489y (1968)
10) – – – Kabachnik, M.I.: Dokl. Akad. Nauk SSSR *182*, 337 (1968); U.S.S.R. Pat. 209 455 (1968); C.A. *69*, 77469s (1968)
11) – – – – Bliznyuk, N.K.: U.S.S.R. Pat. 210 155; C.A. *69*, 59 369e (1968)
12) – Gorelenko, S.V.: Zh. Obshch. Khim. *39*, 836 (1969); E 799
13) Bennett, F.W., Emeléus, H.J., Haszeldine, R.N.: J.Chem.Soc. *1953*, 1565
14) Berlé, F.: J.Prakt. Chem. *66*, 73 (1855); Liebigs Ann. *97*, 334 (1856)
15) Berthaud, J.: Compt. Rend. *143*, 1166 (1906)
16) Bliznyuk, N.K., Kvasha, Z.N., Kolomiets, A.F.: Zh. Obshch. Khim. *37*, 890 (1967), E 840; U.S.S.R. Pat. 179 316 (1966); C.A. *65*, 297h (1966)
17) –, Protasova, L.D., Kvasha, Z.N.: U.S.S.R. Pat. 259 880 (1969); C.A. *73*, 14 987w (1970)
18) – – – Varshavskii, S.L.: U.S.S.R. Pat. 250 134 (1969); C.A. *72*, 79 229z (1970)
19) – – – – Baranov, Yu.I.: U.S.S.R. Pat. 248 675 (1969); C.A. *72*, 79 219w (1970)
20) – – – – U.S.S.R. Pat. 268 418 (1970); C.A. *73*, 88 022z (1970)
21) Bogolyubuv, G.M., Petrov, A.A.: Zh Obshch. Khim. *36*, 1505 (1966); C.A. *66*, 10 995y (1967); Dokl. Akad. Nauk SSSR *173*, 1076 (1967); C.A. *67*, 90 887e (1967)
22) Boyle, R.: New Experiments and Observations Made Upon the Icy Noctiluca, London 1681/2
23) Brago, I.N., Tomilov, A.P.: Elektrokhimiya *4*, 697 (1968); C.A. *69*, 82 856v (1968)
24) Buckler, S.A., Epstein, M.: J.Am.Chem.Soc. *82*, 2076 (1960)
25) Burg, A.B., Mahler, W., Bilbo, A.J., Haber, C.P., Herring, D.L.: J.Am.Chem.Soc. *79*, 247 (1957)
26) Cahours, A., Hofmann, A.W.: Liebigs Ann. *104*, 1 (1857)
27) Carius, L.: Liebigs Ann. *137*, 117 (1866)
28) Carr, R.L.K., Baranauckas, C.F.: U.S.Pat. 3 316 293 (1967); C.A. *67*, 43 921b (1967)
29) Chakravarti, S.N., Dhar, N.R.: J.Phys.Chem. *35*, 1114 (1931)
30) Cohen, S.C., Reddy, M.L.N., Massey, A.G.: Chem.Commun. *1967*, 451
31) Cowley, A.H., Furtsch, T.A., Dierdorf, D.S.: Chem.Commun. *1970*, 523
32) –, Pinnell, R.P.: Inorg. Chem. *5*, 1463 (1966)
33) Crafts, J.M., Silva, R.: J.Chem. Soc. *24*, 629 (1871)
34) Cummins, R.W.: Chem. Ind. (London) *1967*, 918; U.S. Pat. 3 490 940 (1970); C.A. *72*, 100 880x (1970)
35) Deacon, G.B., Parrott, J.C.: J. Organometal. Chem. *17*, P 17 (1969); *22*, 287 (1970)
36) Derkach, N.Ya., Krisanov, A.V.: Zh. Obshch. Khim. *38*, 331 (1968);E 332 (E=Engl.transl.)
37) Dörken, C.: Chem. Ber. *21*, 1505 (1888)

References

38) Eckert, R., Hunger, K., Tavs, P.: Chem. Ber. *100*, 639 (1967)
39) Eméleus, J.H., Smith, J.H.: J. Chem. Soc. *1959*, 375
40) Ernsberger, M.L., Hull, J.W.: U.S.Pat. 2 661 364 (1953)
41) Evers, E.C.: J. Am. Chem. Soc. *73*, 2038 (1951)
42) – Street, E.H., Jr., Jung, S.L.: J. Am. Chem. Soc. *73*, 5088 (1951)
43) Feshchenko, N.G., Alekseeva, T.I., Irodionova, L.F., Kirsanov, A.V.: Zh. Obshch. Khim. *37*, 473 (1967); E 441
44) – – Kirsanov, A.V.: Zh Obshch. Khim. *33*, 1013 (1963); E 1002
45) – – – Zh. Obshch. Khim. *38*, 122 (1968); E 121
46) – – – Zh. Obshch. Khim. *38*, 545 (1968); E 333
47) – Irodionova, L.F., Kirsanov, A.V.: U.S.S.R. Pat. 254 509 (1969); C.A. *72*, 132 970n (1970)
48) – – Korol, O.I., Kirsanov, A.V.: Zh. Obshch. Khim. *40*, 773 (1970); C.A. *73*, 25 575m (1970)
49) – Kirsanov, A.V.: Zh. Obshch. Khim. *36*, 564 (1966); English translation p. 584
50) – Mazepa, I.K., Gorbatenko, Zh.K., Makovetskii. Yu.P., Kukhar', V.P., Krisanov, A.V.: Zh. Obshch. Khim. *39*, 1219 (1969); E 1189
51) – – Makovetskii, Y u. P., Kirsanov, A.V.: Zh. Obshch Khim. *39*, 1886 (1969); E 1846; U.S.S.R. Pat. 260 779 (1970); C.A. *73*, 25 657g (1970)
52) – Mel'nichuk, E.A., Kirsanov, A.V.: Zh. Obshch. Khim. *39*, 2139 (1969); E 2090
53) Filimonova, L.F., Kaabak, L.V., Tomilov, A.P.: Zh. Obshch. Khim. *39*, 2174 (1969)
54) Fluck, E.: Z. Naturforsch. *18b*, 664 (1963)
55) Foley, W.M.: U.S. Pat. 3 150 176 (1964); C.A. *61*, 14 537e (1964)
56) Frank, A.W., Baranauckas, C.F.: J. Org. Chem. *31*, 1644 (1966)
57) Garwood, W.E., Hamilton, L.A., Seger, F.M.: Ind. Eng. Chem. *52*, 401 (1960); U.S. Pat. 2 731 458 (1956)
58) Gee, W., Shaw, R.A., Smith, B.C.: J. Chem. Soc. *1965*, 3354
59) Gladshtein, B.M., Shitov, L.N., Kovalev, B.G., Soborosjii, L.Z.: Zh. Obshch. Khim. *35*, 1570 (1965)
60) – – Zh. Obshch. Khim. *37*, 2586 (1967); E 2461
61) – Soborovskii, L.Z.: U.S.S.R. Pat. 130 513 (1960); C.A. *55*, 6375i (1960)
62) Gmelins Handbuch der anorganischen Chemie, Phosphor, Teil B, System-Number 16, Weinheim: Verlag Chemie 1964
63) Gordon, I., Miller, G.T., Steingart, J.: U.S. Pat. 3 109 785 through 3 109 795 (1963); C.A. *60*, 2552b (1964); see also Ger. Pat. 1 112 722; 1 210 426; 1 210 424; 1 210 425; Brit. Pat. 1 042 391; U.S. Pat. 2 262 871 (1966)
64) Goren, S., Kellerman, D., Greidinger, D.S.: Brit. Pat. 1 075 206 (1967); C.A. *68*, 21 520p (1968)
65) Graham, T.: Quart. J. Sci. *11*, 83 (1829)
66) Grayson, M., Bartkus, E.A.: Pure Appl. Chem. *9*, 193 (1964), ref. 23
67) Guthrie, D.A., Knapp, C.L., Jr.: U.S. Pat. 3 003 964 (1954); C.A. *56*, 7604g (1962)
68) Grohn, H., Friederich, H., Paudert, R.: Z. Chem. *2*, 24 (1962); C.A. *57*, 9701d (1962)
69) Heinrich, R.L.: U.S. Pat. 2 540 580 (1951); C.A. *45*, 5399h (1951)
70) Hettinger, W.P., Jr.: U.S. Pat. 3 079 311 (1963); C.A. *59*, 2859f (1963)
71) Hoffmann, H., Grünewald, R., Horner, L.: Chem. Ber. *93*, 861 (1960)
72) Hofmann, A.W.: Liebigs Ann., Suppl. *1*, 1 (1861)
73) Hurst, T.L.: In: Phosphorus and its Compounds, Vol. II, p. 1149; ed. J.R. Van Wazer. New York: Interscience Publ. Inc. 1961

74) Kaabak, L.V., Kabachnik, M.I., Tomilov, A.P., Varshavskii, S.L.: Zh. Obshch. Khim. *36*, 2060 (1966)

75) — Shandrinov, N.Ya., Tomilov, A.P.: Zh. Obshch. Khim. *40*, 584 (1970); C.A. *73*, 31 026z (1970)

76) Kellerman, D., Wardi, R., Bernstein, H.: Brit. Pat. 1 112 976 (1968); C.A. *69*, 59 365a (1968)

77) Kirsanov, A.V., Fedorova, G.K.: Zh. Obshch. Khim. *37*, 959 (1967); E 908

78) Knapsack, A.G.: Fr. Pat. 1 547 575 (1968); C.A. *71*, 61 546v (1969)

79) Kozlov, N.S., Soshin, V.A.: Tr. Perm. Goz. Sel. Inst. *38*, 83 (1967); C.A. *69*, 96 143x (1968)

80) — — Uch. Zap. Permsk. Gos. Ped. Inst. *32*, 84 (1965); C.A. *66*, 2290k (1967); Zh. Organ. Khim. *2*, 1267 (1966); C.A. *66*, 46 155v (1967); Tr. Perm. Goz. Sed. Inst. *38*, 95 (1967); C.A. *69*, 96 104k (1968)

81) Krafft, F., Neumann, R.: Chem. Ber. *34*, 565 (1901)

82) Kraft, M.Ya., Parini, V.P.: Dokl. Akad. Nauk SSSR *77*, 57 (1951); C.A. *45*, 5478i (1951)

83) — — Sb. Statei Obshchei Khim., Akad. Nauk SSSR *1*, 716 (1953); C.A. *49*, 907i (1955)

84) Krebs, H.: Z. Anorg. Allgem. Chem. *266*, 175 (1951)

85) Kremlev, M.M., Baranovskaya, V.F.: Khim. Tekhnol., Respub. Mezhvedom Nauch.-Tekh. Sb. *1967*, 61; C.A. *69*, 51 783g (1968)

86) Krespan, C.G.: U.S. Pat. 2 996 527 (1961); C.A. *56*, 1483a (1962)

87) — Langkammerer, C.M.: J. Org. Chem. *27*, 3584 (1962)

88) — McKusick, B.C., Cairns, T.L.: J.Am. Chem. Soc. *82*, 1515 (1960). — Krespan, C.G.: J. Am. Chem. Soc. *83*, 3432 (1961)

89) Levchenko, E.S., Piven, Yu. V., Kirsanov, A.V.: Zh. Obshch. Khim. *30*, 1976 (1960); C.A. *55*, 6418i (1961)

90) Lindsell, W.E., Ginsberg, A.P.: 160th ACS-Meeting, Chicago, Sept. 1970, Abstracts of Papers, INORG. 137

91) Maier, L.: Angew. Chem. *77*, 549 (1965); Int. Ed. *4*, 527 (1965)

92) — Chimia *23*, 323 (1969)

93) —Ger. Pat. 1 244 180 (1967); C.A. *67*, 99 629e (1967); U.S. Pat. 3 439 032 (1969)

94) — Helv. Chim. Acta *46*, 2026 (1963); Angew. Chem. *71*, 574 (1959); U.S. Pat. 3 057 917 (1962)

95) — Helv. Chim. Acta *49*, 2458 (1966); U.S. Pat. 3 432 559 (1969)

96) — Helv. Chim. Acta *50*, 1742 (1967)

97) — Helv. Chim. Acta *50*, 1723 (1967); U.S.Pat. 3 359 266 (1967)

98) — Helv. Chim. Acta *51*, 1608 (1968)

99) — Inorganic Syntheses, Vol. VII, p. 82 (1963)

100) — In: Organic Phosphorus Compounds. Kosolapoff, G.M., Maier, L., ed. New York: J. Wiley & Sons, in press. See also Progr. Inorg. Chem. *5*, 27 (1963)

101) — Progr. Inorg. Chem. *5*, 27 (1963)

102) — unpublished

103) Masson, O., Kirkland, J.B.: J. Chem. Soc. *55*, 135 (1889)

104) Märkl, G., Lieb, F.: Angew. Chem. *80*, 702 (1968); Int. Ed. *7*, 733 (1968)

105) McLeod, G.D.: U.S.Pat. 2 768 194 (1956); C.A. *51*, 4699d (1957); C.A. *52*, 9191d (1958)

106) Michaelis, A., Pitch, M.: Chem. Ber. *32*, 337 (1899); Liebigs Ann. *310*, 45 (1899).— Chapman, D.L., Lindbury, A.F.: J. Chem. Soc. *75*, 973 (1899).— Michaelis, A., Arend, K., von: Liebigs Ann. *314*, 259 (1901).— Burgess, C.H., Chapman, D.L.: J. Chem. Soc. *79*, 1235 (1901)

References

107) Miller, C.O.: U.S. Pat. 3 070 581 (1962); C.A. *58*, 11 158d (1963)
108) Montignie, E.: Bull. Soc. Chim. France *49*, 73 (1931)
a) Morgenstern, J., Mayer, R.: Z. Chem. *10*, 449 (1970)
109) Neuworth, M.B., Laufer, R.J.: U.S. Pat. 3 337 637 (1967); C.A. *68*, 12 698e (1968)
110) Okamoto, Y., Sakurai, H.: Kogyo Kagaku Zasshi *68*, 2084 (1965); C.A. *64*, 14 210g (1966)
111) – – Kogyo Kagaku Zasshi *69*, 1557 (1966); C.A. *66*, 37 576f (1967)
112) Oppegard, A.L.: U.S. Pat. 2 687 437 (1954); C.A. *49*, 11 000h (1955)
113) Osadchenko, I.M., Tomilov, A.P.: Zh. Obshch. Khim. *39*, 469 (1969)
114) – – Zh. Obshch. Khim. *40*, 698 (1970); C.A. *73*, 20 834k (1970); U.S.S.R. Pat. 256 762 (1969); C.A. *72*, 132 951g (1970)
115) Parini, V.P., Kraft, M.Ya.: Sb. Statei Obshchei Khim., Akad. Nauk. SSSR *1*, 723 (1953); C.A. *49*, 909d (1955)
116) – – Sb. Statei Obshch. Khim., Akad. Nauk. SSSR *1*, 729 (1953); C.A. *49*, 909e (1955)
117) Perner, D., Henglein, A.: Z. Naturforsch. *17b*, 703 (1962); see also Radiation Chem. Proc. *1962*, Tihany Symp., Hung., *1962*, 427; C.A. *62*, 7286a (1965)
118) Peterson, D.J., Logan, T.J.: J. Inorg. Nucl. Chem. *28*, 53 (1966)
119) Petrov, K.A., Smirnov, V.V., Emel 'yanov, V.I.: Zh. Obshch. Khim. *31*, 3027 (1961); E 2823
120) Praetzel, H.E., Jenkner, H.: DAS 1 232 578 (1967); DAS 1 249 274 (1967)
121) Procter & Gamble Co.: Neth. Appl. 6 610 762 (1967); C.A. *67*, 108 751t (1967)
122) Protasova, L.D., Kvasha, Z.N., Bliznyuk, N.K., Varshavskii, S.L., Baranov, Yu.I.: U.S.S.R. Pat. 258 309 (1969); C.A. *72*, 13 954k (1970)
123) Rankov, G.: Ann. Univ. Sofia, II, Fac. Phys. Math., Livre *2*, 32, 51–102 (1936); C.A. *31*, 2168[5] (1937)
124) Rauhut, M.M.: Topics in Phosphorus Chemistry, Vol. *1*, 1 (1964)
125) – Semsel, A.M.: J. Org. Chem. *28*, 473 (1963); U.S.Pat. 3 099 691 (1963); C.A. *60*, 555 (1964)
126) – Bernheimer, R., Semsel, A.M.: J. Org. Chem. *28*, 478 (1963)
127) – Semsel, A.M.: U.S. Pat. 3 099 684 (1963); C.A. *60*, 555g (1964)
128) – – J. Org. Chem. *28*, 471 (1963); DAS 1 197 449 (1965); U.S. Pat. 3 060 241 (1962); C.A. *58*, 6862c (1963)
129) – – U.S. Pat. 3 099 690 (1963); C.A. *60*, 556a (1964)
130) Ritchey, H.W.: U.S. Pat. 2 311 305 (1943); C.A. *37*, 4891[6] (1943)
131) Rio, A.: Ger. Offen. 1 945 645 (1970); C.A. *72*, 132 955m (1970); Fr. Pat. 1 584 799 (1970)
132) Rochow, E.G.: J. Am. Chem. Soc. *67*, 963 (1945)
133) Scheffler, M., Drawe, H., Henglein, A.: Z. Naturforsch. *23b*, 911 (1968)
134) – Henglein, A.: Z. Naturforsch. *25b*, 103 (1970)
135) Schumann, H., Köpf, H., Schmidt, M.: Z. Anorg. Allgem. Chem. *331*, 200 (1964)
136) – – – Chem. Ber. *97*, 1458 (1964);
a) Schumann, H.: private communication
137) Senear, A.E., Valient, W., Wirth, J.: J. Org. Chem. *25*, 2001 (1960)
138) Shandrianov, V.Ya., Tomilov, A.P.: Elektrokhimiya *4*, 237 (1968)
139) Steinkopf, W., Buchheim, K.: Chem. Ber. *54*, 1024 (1921)
140) Stevens, D.R., Spindt, R.S.: U.S. Pat. 2 542 370 (1951); C.A. *45*, 5712g (1951)
141) Street, E.H., Jr., Gardner, D.M., Evers, E.C.: J.Am. Chem. Soc. *80*, 1819 (1958)
142) Stuebe, C., LeSuer, W.M., Norman, G.R.: J.Am. Chem. Soc. *77*, 3526 (1955)
143) Thenard, R.: Compt. Rend. *21*, 144 (1845); *25*, 892 (1847); Jahresber. *1847 – 1848*, 645

58

144) Titov, A.I., Gitel, P.O.: Dokl. Akad. Nauk SSSR *158*, 1380 (1964); E 1119
145) Union Carbide Corp.: Belg. Pat. 730 091 (1969)
146) Van Wazer, J.R.: Phosphorus and its Compounds, Vol. 1. New York: Interscience Publishers Inc. 1958
147) Varshavskii, S.L., Kaabak, L.V., Kabachnik, M.I., Tomilov, A.P.: U.S.S.R. Pat. 222 382 (1968); C.A. *70*, 4271g (1969)
148) − Tomilov, A.P., Smirnov, Y.u.D.: Zh. Vses. Khim., Obshchestva im. D.I. Mendeleeva *7*, 598 (1962); C.A. *58*, 3097h (1963)
149) Vol 'fkovich, S.I., Kuskov, V.K., Koroteeva, K.F.: Izv. Akad. Nauk SSSR, Otdel. Khim. Nauk *1954*, 5; C.A. *49*, 6859c (1955)
150) Walling, C., Stacey, F.R., Jamison, S.E., Huyser, E.S.: J.Am. Chem. Soc. *80*, 4543 (1958)
151) − − − − J.Am. Chem. Soc. *80*, 4546 (1958)
152) Watson, W.H.: Texas J. Sci. *11*, 471 (1959); C.A. *54*, 13 928c (1960)
153) Willstätter, R., Sonnenfeld, E.: Chem. Ber. *47*, 2801 (1914); U.S.Pat. 1 205 138 (1916)
154) Wu, C.: J.Am. Chem. Soc. *87*, 2522 (1965)
155) − U.S. Pat. 3 420 917 (1969); C.A. *70*, 58 018y (1969)
156) − U.S. Pat. 3 458 581 (1969); C.A. *71*, 102 007r (1969)
157) Wunsch, G., Wintersberger, K., Geierhaas, H.: Z. Anorg. Allgem. Chem. *369*, 33 (1969)
158) Zabolotny, E.R., Gesser, H.: J.Am. Chem. Soc. *81*, 6091 (1959)
159) Zhuravleva, L.P., Z'ola, M.I., Suleimanova, M.G., Kirsanov, A.V.: Zh. Obshch. Khim. *38*, 342 (1968); E 341

Received January 1971

Studies in Phosphorus Stereochemistry

Dr. G. Zon* and Prof. Dr. K. Mislow**

Department of Chemistry, Princeton University, Princeton, N.J. 08540, USA

Contents

* Public Health Service Predoctoral Fellow, 1969-71.

** To whom correspondence should be addressed.

1. Introduction

The present review summarizes recent studies carried out at Princeton which have dealt with the stereochemistry of displacement reactions at phosphorus. The review is divided into two principal sections. The first part (Sect. 2) describes primarily the development of general synthetic methods for the stereospecific conversion of phosphinates → phosphine oxides → phosphines, and the configurational intercorrelations of these optically active organophosphorus compounds. The second part (Sect. 3) evaluates the role of pseudorotation in the stereochemistry of various displacement reactions at tetracoordinate phosphorus. The influence of steric and electronic factors is discussed with the aid of a general topological representation which maps the stereochemistry of displacement reactions at tetracoordinate centers.

2. Synthesis and Configurational Correlation of Phosphinates, Phosphine Oxides, and Phosphines

2. 1. Stereospecific Conversion of Menthyl[a] Phosphinates to Optically Active Phosphine Oxides Using Organometallic Reagents

Optically active phosphine oxides occupy a central position in the study of organophosphorus reaction mechanisms and stereochemistry [1]. Because the reported synthetic approaches [1] to these oxides lacked flexibility and were of limited use in configurational intercorrelations, a stereochemical investigation of the reaction of Grignard reagents with menthyl phosphinates was initiated [2], conceptually analogous to Andersen's Grignard synthesis of optically active sulfoxides from menthyl sulfinates [3]. Chart I illustrates how the configurations of trialkyl-, dialkylaryl-, alkyldiaryl-, and triarylphosphine oxides may in principle be correlated *via* the appropriately substituted phosphinates, assuming that the Grignard reaction maintains the same stereochemical direction throughout.

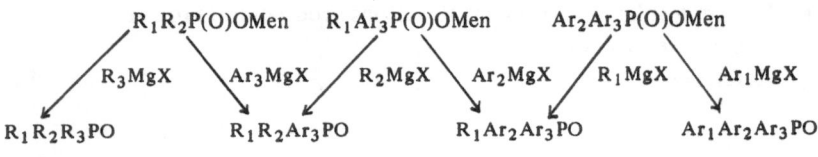

Chart I

a) Throughout this review, "menthyl" (= OMen) refers to that group derived from natural (-)-menthol.

Requisite diastereomerically enriched menthyl phosphinates may be conveniently obtained by unexceptional methods [2], and examples which constitute the partial realization of this scheme for configurational intercorrelations are shown in Chart II. The absolute configurations of menthyl esters *1* and *2* follow from their correlation with *3*, whose chirality at phosphorus is known from X-ray analysis [2]. The stereochemical direction (inversion) of these highly stereospecific Grignard reactions, and thus the absolute configurations of the derived optically active phosphine oxides, was established by chemical correlations with a second reference compound *(4)* of known absolute configuration [4].

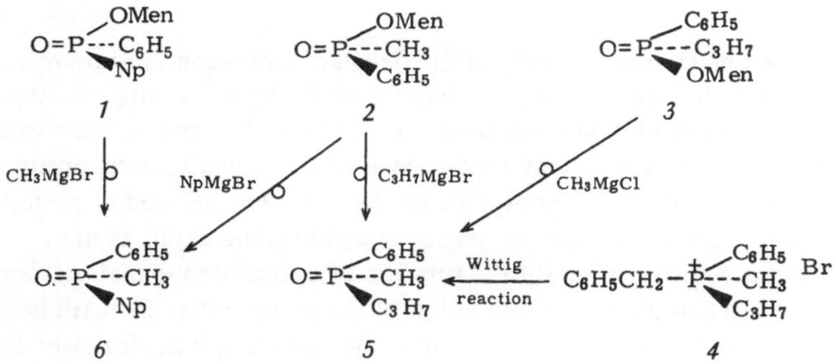

Chart II

Np = β-naphthyl

The extreme sensitivity of the Grignard synthesis to variations in the groups on either phosphorus or on magnesium imposes a serious limitation of its scope and foiled early attempts [2] to realize all of the intercorrelations in Chart I. Accordingly, two modifications were investigated [5]: a structural modification of the phosphinate, to reduce the steric requirement at phosphorus during nucleophilic substitution, and use of the more reactive lithium reagents, which have been employed in similar reactions in cases where Grignard reagents have failed [6]. These changes proved successful and the results are summarized in Chart III.

However, in contrast to the Grignard reactions of phosphinates, substitution with organolithium reagents was found to be sometimes significantly less stereospecific. Such loss of overall stereospecificity may be ascribed [5] to one or a combination of factors: epimerization of the starting phosphinate, stereomutation of pentacoordinate intermediates by intramolecular ligand exchange (see Sect. 3), or partial racemization of product through exchange [7] of groups bonded to phosphorus.

$$O=P{\small\begin{matrix}\diagup OMen\\ ---C_2H_5\\ \diagdown CH_3\end{matrix}}$$

```
        1                        2                              O=P---C_2H_5
                                                                   CH3
      AnLi        CH3MgBr              C2H5 MgBr        C6H5MgBr      C3H7MgBr
                  or CH3Li    NpMgBr   or C2H5Li
                                                                        C3H7
        Np                              C6H5                    O=P---CH3
  O=P---C6H5               6       O=P---CH3                       C2H5
     An                                 C2H5
        7
```

Chart III

Np = β-naphthyl; An = o-anisyl

To assess the stereospecificity of the Grignard and organolithium reactions with menthyl phosphinates, the diastereomeric purity of starting menthyl esters was estimated by pmr spectroscopy (see Sect. 2.2) and, in most cases, highest reported rotations were used to estimate the enantiomeric purity of the derived optically active phosphine oxides [2, 5]. The method of preference for determining the enantiomeric purity of a phosphine oxide, even in those cases in which a value for the rotation of optically pure material is reported, involves stereospecific reduction of the phosphine oxide with hexachlorodisilane (see Sect. 2.4) to the corresponding phosphine, followed by quaternization with 2-phenyl-2-methoxy-ethyl bromide and pmr analysis of the diastereomeric phosphonium bromides (Eq. (1)) [8, 9]. This method for determining optical purity, shown [8] to be applicable

$$\overset{*}{R_3P} + BrCH_2CH(OCH_3)C_6H_5 \longrightarrow \overset{+*}{R_3P} - CH_2\overset{*}{CH}(OCH_3)C_6H_5 \ Br^- \quad (1)$$

to aliphatic, aromatic, and mixed tertiary phosphines, has allowed assessment of the absolute rotation of 7, and hence the stereospecificity of the reaction of diastereomerically pure 1 with o-anisyllithium [5].

2.2. Configurational Correlation of Menthyl Phosphinates by Nuclear Magnetic Resonance

Menthyl phosphinates, in addition to being useful precursors for the synthesis of optically active tertiary phosphine oxides [2, 5], phosphinamides [10], and phosphonothioates [11], were found [12, 13] to exhibit pmr spectra which are a rich lode of structural information. For menthyl n-alkylphenylphosphinates (8) the upfield portion of the pmr spectra features the three methyl doublets of the menthyl moiety. Chemical shifts of these signals for diastereomers

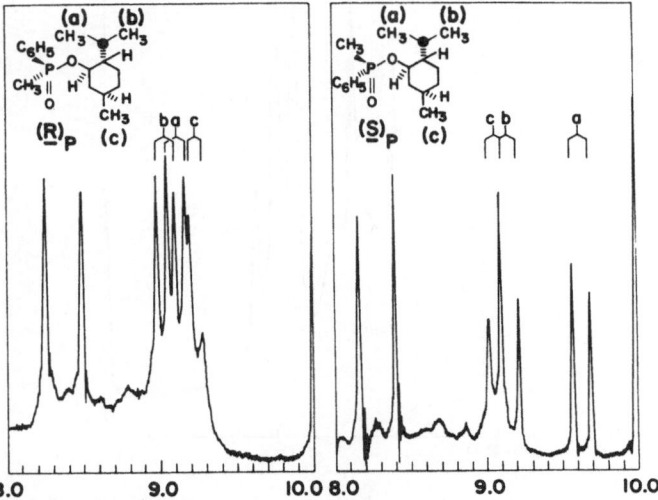

8

which differ in chirality at phosphorus are anisochronous and may thus be conveniently employed as monitors of diastereomeric purity. For example, the upfield position of the H_a doublet in *(S)*p-2 (Fig. 1), which is in a region

Fig. 1. Pmr spectra of diastereomeric menthyl methylphenylphosphinates *(R)*p-2 (left) and *(S)*p-2 (right), τ scale

unencumbered by other signals, readily allows detection and quantitative estimation of any significant contamination of *(R)*p-2 by *(S)*p-2. Furthermore, since a similar upfield resonance of H_a was observed in *(S)*p-3 but not in *(R)*p-3, the position of H_a was suggested as diagnostic of phosphorus chirality: an upfield shift corresponding to the *(S)*p configuration [13].

As a strictly empirical configurational correlation, applicable to compounds belonging to the system represented by *8*, it is not necessary to identify the source of the observed upfield shift, or the identity of the diastereotopic [14] isopropyl methyl group which exhibits this diagnostic resonance. However, in order to gain further insight into these phenomena, and in order to provide a basis for extensions to cognate systems, the diastereotopic methyl groups were identified, by the following procedure.

65

The large upfield shift (*ca.* 0.5 ppm) of the H_a doublet in *(S)*P-*2* and *(S)*P-*3* was readily attributed to the diamagnetic anisotropy of the phenyl ring, rather than the phosphoryl group, for this shift was not observed when the phenyl group in *2* was replaced with a cyclohexyl group [13] (Fig. 2).

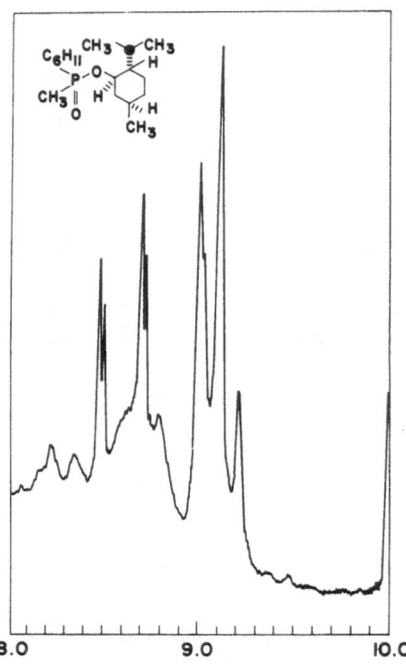

Fig. 2. Pmr spectrum of a mixture of diastereomeric menthyl cyclohexylmethylphosphinates, τ scale

Although anisochrony of diastereotopic isopropyl methyl groups is frequently encountered, the rigorous identification of such groups had been reported [15] only once, prior to the study under discussion. As in the previous investigation [15] this identification procedure utilized the pmr spectra of isotopically substituted derivatives in which the methyl groups under consideration had been stereospecifically deuterated. Chart IV summarizes the synthetic route used [13] to prepare the two appropriately deuterated (-)- menthols: *11*, in which the *pro-S* [16] methyl group had been replaced by a deuteriomethyl group, and *14*, in which the *pro-R* methyl group had been replaced by a trideuteriomethyl group. The key step in this synthetic scheme was the oxidative hydroboration of *9* and *12*, an asymmetric synthesis which proceeds with a high degree of stereospecificity [17] to give *10* [13] and *13*, respectively.

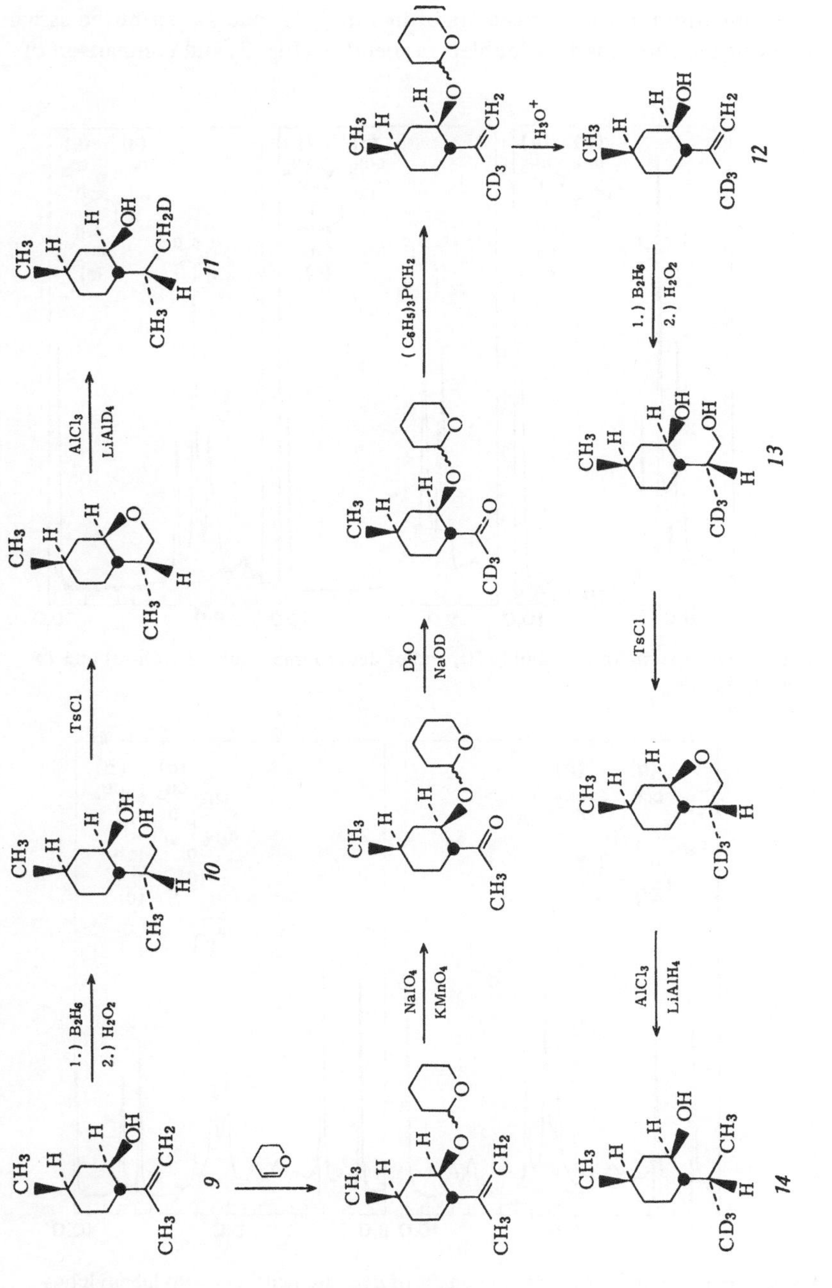

Chart IV

67

Comparison of the pmr spectra of menthol, *11*, and *14* permitted assignment of the three methyl doublets in menthol (Fig. 3) and comparison of

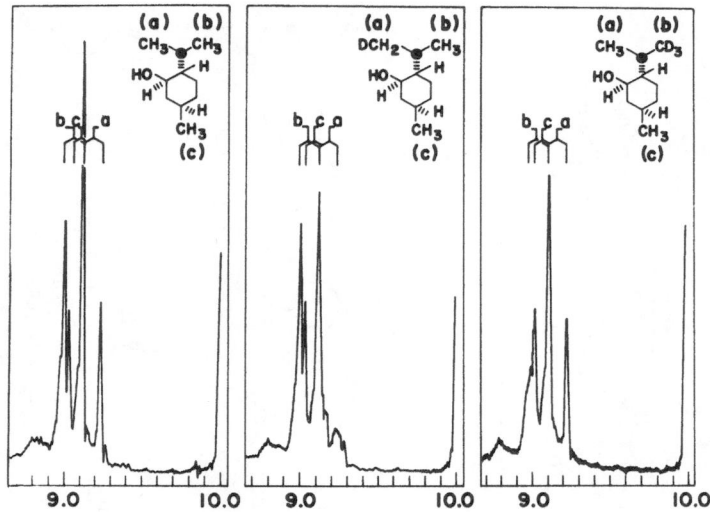

Fig. 3 Pmr spectra of menthol (left), and of deuteriomenthols *11* (center) and *14* (right), *τ* scale

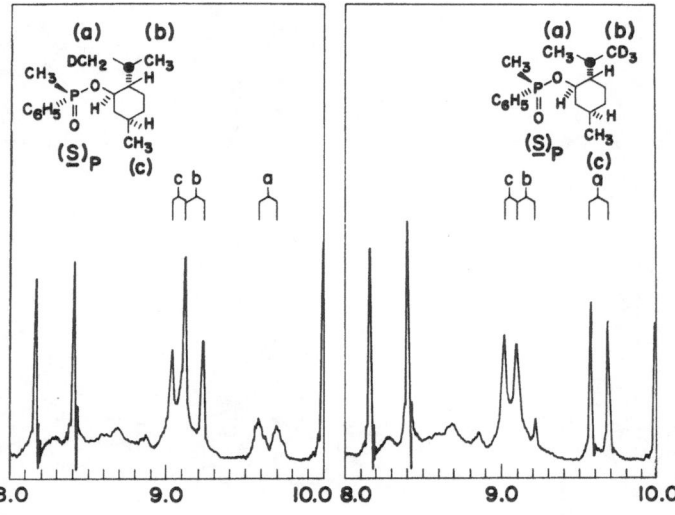

Fig. 4. Pmr spectra of (*S*)p diastereomers of deuteriomenthyl methylphenylphosphinates prepared from *11* (left) and *14* (right), *τ* scale

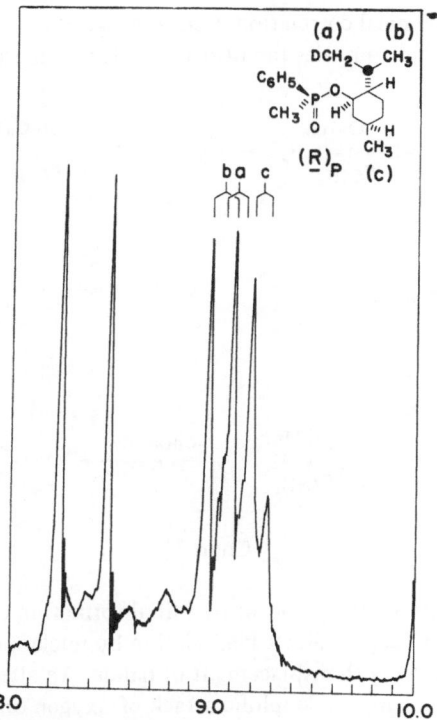

Fig. 5.   Pmr spectrum of the (R)p diastereomer of deuteriomenthyl methylphenyl-phosphinate prepared from *11*, τ scale

the pmr spectra of the deuterated diastereomers of *2* (Figs. 4 and 5) with those of ordinary *2* (Fig. 1) led to the unambiguous identification of the upfield shifted protons *(H_a)* as those of the *pro-S* methyl group [13].

Configurational assignments based on the above pmr correlation have been made for menthyl benzylphenylphosphinate [18], phenylphosphinate [19], and isopropylphenylphosphinate [19]. Finally, a complete parallelism in pmr characteristics has been shown to exist between phosphinates *8* and menthyl *p*-iodobenzenesulfinate [20].

2.3. Stereospecific Alkylation of Menthyl Phenylphosphinate

A synthetically useful development in the stereochemistry of menthyl phosphinates has been achieved by the finding that menthyl phenylphosphinate may be stereospecifically alkylated with retention of configuration [19]. Thus, a single precursor suffices for the synthesis of a wide variety of alkyldiaryl- and aryldialkylphosphine oxides of known absolute configuration. Chart V

69

summarizes the chemical correlation used to assign the stereochemistry of the alkylation, and exemplifies the utility of this new method.

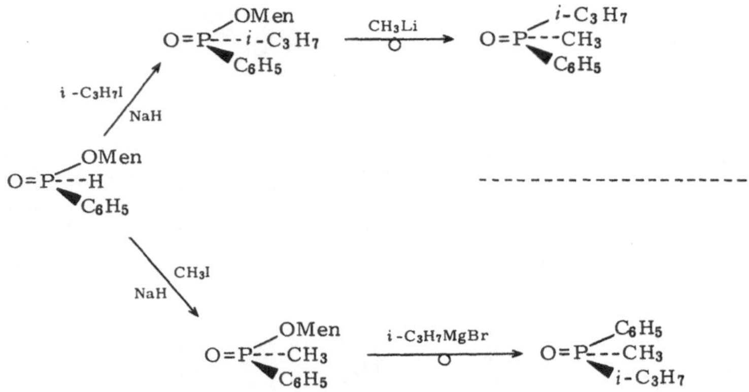

Chart V

The reaction is presumed to occur by initial formation of $\lfloor C_6H_5(MenO)PO\rfloor^-$, which then undergoes direct P-alkylation by nucleophilic attack of phosphorus on carbon with displacement of halide. An alternative alkylation mechanism involving nucleophilic attack of oxygen on carbon, followed by a Michaelis-Arbuzov rearrangement [21] of a dialkyl phenylphosphonite $(C_6H_5(MenO)POR)$ intermediate with the alkyl halide was effectively eliminated by the observation [19] that reaction of methyl phenylphosphinate with a tenfold excess of methyl-d_3 iodide gave the product distribution shown in Eq. (2).

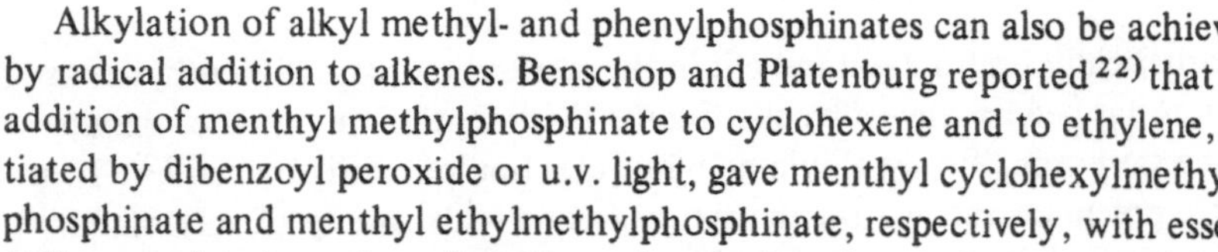

Alkylation of alkyl methyl- and phenylphosphinates can also be achieved by radical addition to alkenes. Benschop and Platenburg reported [22] that addition of menthyl methylphosphinate to cyclohexene and to ethylene, initiated by dibenzoyl peroxide or u.v. light, gave menthyl cyclohexylmethylphosphinate and menthyl ethylmethylphosphinate, respectively, with essentially complete inversion of configuration at phosphorus. In contradistinction, we have found that dibenzoyl peroxide-catalyzed addition of menthyl phenylphosphinate to cyclohexene occurs with retention of configuration at : phosphorus [23]. This apparent discrepancy was resolved, as follows [24]. Alkylation of menthyl methylphosphinate with ethyl iodide/sodium hydride gave the same diastereomer of menthyl ethylmethylphosphinate which resulted from the photochemical addition of ethylene to menthyl methylphosphinate (Chart VI). Given that the former reaction occurs with retention of configuration at phosphorus [19], it follows that free radical additions of menthyl *methyl*phosphinate and menthyl *phenyl*phosphinate to olefins both occur with retention of configuration at phosphorus, and that the tentative assignment of absolute configuration at phosphorus in menthyl methylphosphinate by Meppelder, Benschop, and Kraay [25], on which the earlier claims were based [22], must be in error. As an additional consequence of these findings, it now appears that the conversion of menthyl methylphosphinate to menthyl methylphenylphosphinate *via* menthyl S-methyl methylphosphonothioate (Chart VI), reported by Benschop and Platenburg [22], must proceed

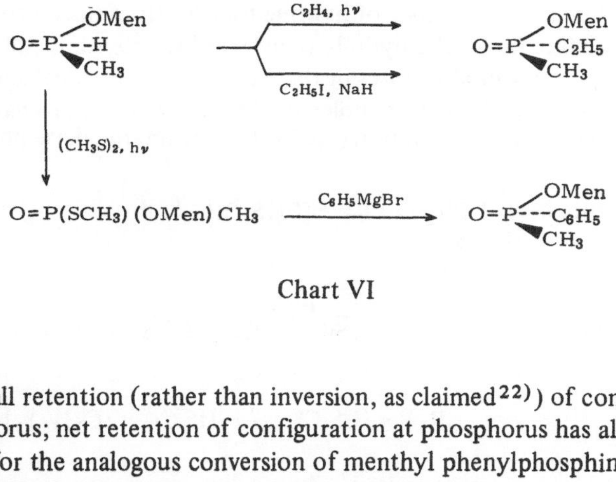

Chart VI

with overall retention (rather than inversion, as claimed [22]) of configuration at phosphorus; net retention of configuration at phosphorus has also been observed for the analogous conversion of menthyl phenylphosphinate to menthyl methylphenylphosphinate *via* menthyl S-methyl phenylphosphonothioate [24]. This means that the photochemical thiomethylation and the Grignard displacement must proceed with the same stereochemistry, *i.e.*, both with retention or both with inversion. Inversion stereochemistry for the thiomethylation or retention stereochemistry for the Grignard displacement is highly unexpected. The latter possibility has considerable import, as it bears upon whether the assigned absolute configurations for sarin and other anticholinesterases [26] are correct.

2.4. Stereospecific Reduction of Phosphine Oxides to Phosphines

Tertiary phosphines, in the absence of special effects [27], have relatively high barriers [28] (*ca.* 30–35 kcal/mol) to pyramidal inversion, and may therefore be prepared in otically stable form. Methods for synthesis of optically active phosphines include cathodic reduction [29], or base-catalyzed hydrolysis [30, 31] of optically active phosphonium salts, reduction of optically active phosphine oxides with silane hydrides [32], and kinetic [31] or direct [33] resolution. The ready availability of optically pure phosphine oxides of known absolute configuration by the Grignard method (see Sect. 2.1) led to a study [34] of a convenient, general, and stereospecific method for their reduction, thus providing a combined methodology for preparation of phosphines of known chirality and of high enantiomeric purity.

Horner and Balzer had earlier reported [32] that reduction of optically active phosphine oxides with either trichlorosilane ($HSiCl_3$), $HSiCl_3$/pyridine, or $HSiCl_3$/N, N-diethylaniline affords phosphines with overall retention of configuration, whereas reduction with $HSiCl_3$/triethylamine affords phosphine with inversion of configuration at phosphorus. In summary, it was suggested [32] that this difference in overall stereochemistry of reduction reflected a difference in the mode of hydride transfer from silicon to phosphorus: intra- and intermolecular hydride transfer led to retention and inversion, respectively. The essential features of these mechanistic rationalizations are represented by Eq. (3). The intramolecular hydride transfer mechanism [32], which may include pseudorotation (see Sect. 3) if intermediate phospho-

$$HSiCl_3 + O=\overset{*}{P}R_3 \rightarrow \overset{+}{O}\overset{\frown}{\underset{|}{\rule{0pt}{1.4ex}}}\overset{+}{P}R_3 \rightarrow \overset{*}{P}R_3 + [Cl_3SiOH]$$
$$Cl_3\underline{Si}\overset{\frown}{\rule{0pt}{1.4ex}}H$$

$$\tag{3}$$

$$HSiCl_3 + O=\overset{*}{P}R_3 \xrightarrow{(C_2H_5)_3N} HCl_3\bar{Si}O\overset{\frown}{\underset{}{\rule{0pt}{1.4ex}}}\overset{+}{P}R_3 \overset{\frown}{\rule{0pt}{1.4ex}} H\overset{-}{\underset{}{\rule{0pt}{1ex}}}\bar{Si}Cl_3 \cdot \overset{+}{N}(C_2H_5)_3$$

$$R_3\overset{*}{P} + (OSiCl_2)_n + HSiCl_3 + (C_2H_5)_3\overset{+}{N}HCl^-$$

ranes ((Cl_3SiO) (H) PR_3) are involved [34], remains consistent with available stereochemical data, and recently a similar mechanism has been proposed for the closely related $HSiCl_3$ reduction of aromatic sulfoxides [35]. However, a more thorough survey [34] of the $HSiCl_3$/tertiary amine systems revealed a striking dependence of the overall stereochemistry of reduction on the nature of the accompanying tertiary amine: strong bases (pK_b < *ca.* 5) gave phosphine with predominant inversion and weak bases (pK_b > *ca.* 7) gave phosphine with predominant retention of configuration. Furthermore, evidence was presented in support of the hypothesis [34] that the inversion of configuration observed in the presence of strongly basic tertiary amines may result from reduction *via* trichlorosilyl anion, as shown in Eq. (4), and/or *via* related perchloropolysilanes (Si_nCl_{2n+2}) or silicon subhalides $((SiCl_2)_n)$, which may result from amine-catalyzed decomposition of $HSiCl_3$. Recent pmr studies [36, 37] of $HSiCl_3$/tertiary amine systems are consistent with the equilibrium shown in the first step of Eq. (4).

$$
\begin{aligned}
HSiCl_3 + R_3N &\rightleftharpoons {}^-SiCl_3 + R_3\overset{+}{N}H \\
R_3\overset{*}{P}{=}O + R_3\overset{+}{N}H &\rightleftharpoons R_3\overset{+}{P}OH + R_3N \\
Cl_3Si^- + R_3\overset{+}{P}OH &\xrightarrow{-\sigma} Cl_3Si-\overset{+}{P}R_3 + {}^-OH \\
HO^- + Cl_3Si-\overset{+}{P}R_3 &\longrightarrow [HOSiCl_3] + \overset{*}{P}R_3
\end{aligned}
\tag{4}
$$

In order to test the above hypothesis, an investigation of the reducing properties of lower members of the perchloropolysilanes [b], *i.e.*, hexachlorodisilane $(n=2)$ and octachlorotrisilane $(n=3)$, was initiated and led to the observation [34, 39] that both of these compounds, and presumably their higher homologues, reduce optically active acyclic phosphine oxides with complete or nearly complete inversion of configuration. The stoichiometry and stereochemistry of the Si_2Cl_6 reduction of acyclic phosphine oxides are satisfactorily, and most simply [34], accounted for by the scheme in Eq. (5), wherein inversion results from nucleophilic attack on phosphorus by trichlorosilyl anion. Details regarding the possibility of intermediate phosphoranes and the effect on the stereochemistry of this displacement during reduction of cyclic phosphine oxides with Si_2Cl_6 are discussed in Sect. 3.2.

$$
R_3\overset{*}{P}{=}O + Si_2Cl_6 \longrightarrow R_3\overset{+}{P}-OSiCl_3 + {}^-SiCl_3 \xrightarrow{-\sigma} Cl_3Si-\overset{+}{P}R_3 + {}^-OSiCl_3
$$

$$
Cl_3SiO^- + Cl_3Si-\overset{+}{P}R_3 \left\}
\begin{aligned}
&\longrightarrow Cl_3SiOSiCl_3 + \overset{*}{P}R_3 \\
&\longrightarrow Cl_4Si + \overset{*}{P}R_3
\end{aligned}
\right.
\tag{5}
$$

$$
\downarrow
$$

$$
[OSiCl_2] + Cl^-
$$

Hexachlorodisilane has been found [34] to be a convenient deoxygenating agent of amine oxides and sulfoxides, and has been used in the synthesis of condensed bridged phosphines [40]. In addition, the first and only stereospecific desulfurization of phosphine sulfides reported to date was also accom-

---

b) Earlier investigations by Urry [38] led to recognition that perchloropolysilanes may function as reducing agents.

plished using Si_2Cl_6 and, in contrast to what was observed with the corresponding phosphine oxides, this desulfurization of acyclic phosphine sulfides was found to proceed with retention of configuration at phosphorus [41]. To rationalize these stereochemical differences, it was suggested[41] that with phosphine sulfides, attack at sulfur by trichlorosilyl anion and elimination of phosphine (retention, Eq. (6)) successfully competes with attack on phosphorus (inversion, Eq. (5)). Stabilization of the intermediate or transition state represented by *15*, due to (p-d)π bonding, which is not likely for the oxygen ana-

$$R_3\overset{*}{P}=S + Si_2Cl_6 \longrightarrow R_3\overset{+}{P}-SSiCl_3 + {}^-SiCl_3 \tag{6}$$

$$R_3\overset{*}{P} + Cl_6Si_2S \longleftarrow \left[R_3P=S \begin{matrix} SiCl_3 \\ SiCl_3 \end{matrix} \right]_{15}$$

logue, may account for this stereochemical dichotomy. It should be noted however, that such stabilization, if present, does not lead to a similar reversal in stereochemistry of displacement for the closely related alkaline hydrolyses shown in Eq. (7), since both ethoxy- (X = O) and ethylmercapto- (X = S) phosphonium ions have been shown [41] to give phosphine oxide with inversion of configuration at phosphorus.

$$R_3\overset{*}{P}=X \xrightarrow{(C_2H_5)_3\overset{+}{O}} R_3\overset{+}{P}-XC_2H_5 \xrightarrow{{}^-OH} O=\overset{*}{P}R_3 \tag{7}$$

3. The Role of Pseudorotation in the Stereochemistry of Displacement Reactions at Tetracoordinate Phosphorus [c]

3.1. Pseudorotation of Pentacoordinate Intermediates and the Utility of Topological Representations

Bimolecular nucleophilic substitution at tetracoordinate phosphorus (Eq. (8)) may proceed by either direct (S_N2) substitution or by an addition-elimination mechanism [d]. In the former, *16* represents a transition state, while in the

---

c) A substantial portion of Sect. 3. has been adapted from Ref.[42].

d) Unimolecular substitution at second or higher row elements is considered to be a relatively less favored [43] mode of reaction.

latter, *16* represents a pentacoordinate (phosphorane) intermediate, which may, or may not be operationally detectable [e].

$$X^- + L_3\overset{+}{P}-Y \rightleftarrows L_3PXY \rightleftarrows L_3\overset{+}{P}-X + Y^- \qquad (8)$$

16

One of the ways in which a transient phosphorane may be detected is by its stereochemical non-rigidity: phosphoranes may undergo intramolecular ligand exchange (polytopal rearrangement [46]) by Berry pseudorotation [47], wherein pairwise exchange of apical (*a*) and equatorial (*e*) ligands in the trigonal-bipyramidal [48] molecule takes place by way of a tetragonal-pyramidal transition state [f]. The "pivot" ligand, which remains equatorial in this process, occupies the apex of the pyramid in the transition state.

Pseudorotation about pivot ligand 1 is illustrated by Eq. (9). To be operationally detectable, it is necessary that the energy barriers for pseudorotation are accessible, and that the phosphorane has a sufficient lifetime, relative to its tetracoordinate reaction partners.

$$(9)$$

As illustrated in Eq. (10), a chiral phosphonium ion can undergo attack by a nucleophile at any one of four different faces or six different edges, thus placing the entering ligand in the *a* and *e* positions, respectively. In the general case, when all five ligands are different, and in the absence of special constraints (see Sect. 3.2) 20 isomeric phosphoranes, which are interconnected by 30 pseudorotation steps, are thus produced from both enantiomers of the phosphonium ion. Because of the possibility for reaction *via* this complex intermediate manifold, interpretation of the stereochemical consequences of

---

e) It has been suggested [44] that comparative rate studies of nucleophilic displacement reactions in the phosphetane ring system with appropriate acyclic models offer criteria which may enable discrimination between associative displacement through a pentacoordinate intermediate and an S_N2-like direkt displacement mechanism in this cyclic system. However, see Ref. [45].

f) Alternative mechanisms are conceivable [49]. However, see Ref. [50].

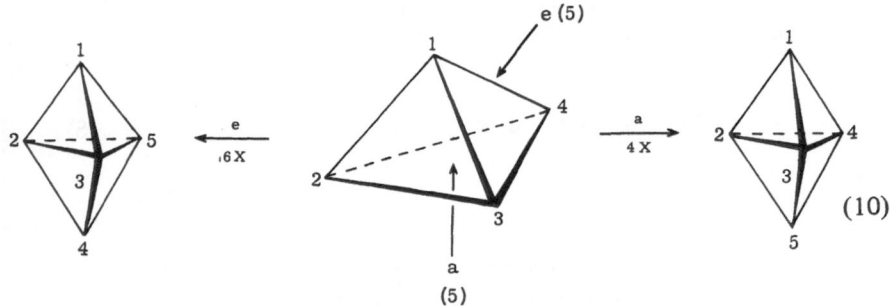

$$(10)$$

substitution at tetracoordinate phosphorus poses a formidable problem. Such an analysis may be accomplished by systematically listing all of the possible intermediates and their pseudorotational interconversions. However, topological representations provide greater convenience and economy [g].

The set of 20 interconnecting isomers and the network of 30 interconnecting pseudorotations may be displayed in the form of a Desargues-Levi graph [h], with isomers and pseudorotations represented by vertices and edges, respectively. The first chemical application of the Desargues-Levi graph was reported by Balaban, *et al.* [53], who applied it to 1,2-shifts of carbonium ions. More recently, others [54] have utilized the same graph [i] to describe pseudorotational interconversions of stereoisomeric phosphoranes. As shown in Fig. 6,

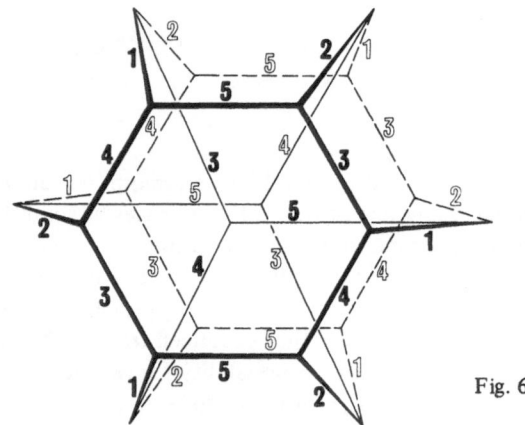

Fig. 6

our [56] own geometric realization of the Desargues-Levi graph has idealized D_{3d} symmetry. Indices of the pivot (nesessarily e) ligands for pseudorotation are designated by the numerals over the edges and are related through the center of symmetry. Consequently, the identity of each isomer is automatically defined and according to the convention employed [56], isomers are designated by the apical ligands. Chirality is denoted, arbitrarily, by the ascending numerical order of equatorial ligand indices for each isomer: if clockwise when viewed from the a ligand with lower numerical index, the isomer is unbarred; if counterclockwise, barred. The isomer designations thus generated are given in Fig. 7.

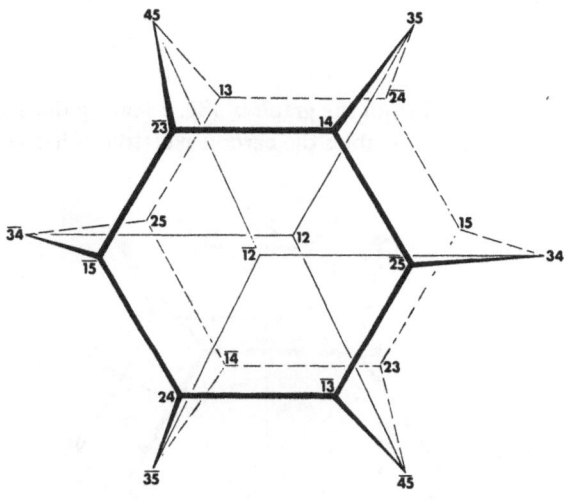

Fig. 7

Inter- and intramolecular displacement reactions at phosphorus which proceed by way of intermediate phosphoranes incorporated in a small-ring system may be generalized by Eq. (11), in which the convention of assigning indices 1 and 2 to the ring termini is employed. Since a small-ring system is incapable of spanning the a positions, diastereomer $12/\overline{12}$ and its six connect-

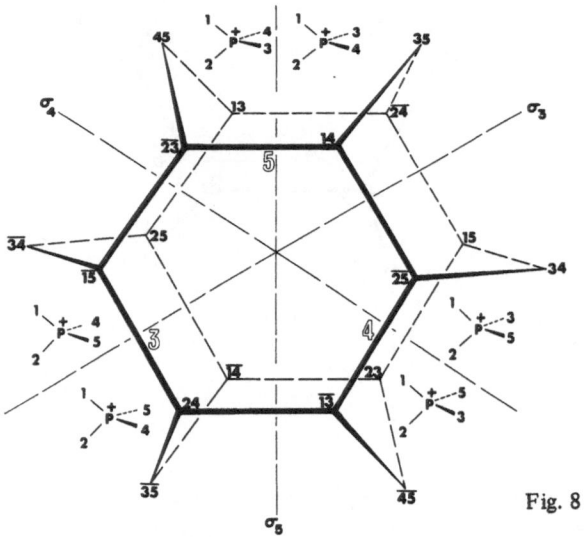

$$L_3^- \quad + \quad \begin{matrix} L_1 & & L_4 \\ & \overset{+}{P} & \\ L_2 & & L_5 \end{matrix} \qquad\qquad L_1^- \quad \begin{matrix} L_2 & & L_3 \\ & \overset{+}{P} & \\ L_5 & & L_4 \end{matrix}$$

$$L_4^- \quad + \quad \begin{matrix} L_1 & & L_3 \\ & \overset{+}{P} & \\ L_2 & & L_5 \end{matrix} \quad\rightleftharpoons\quad \begin{matrix} L_1 & & L_3 \\ & P & \!\!\!-\; L_4 \\ L_2 & & L_5 \end{matrix} \qquad (11)$$

$$L_5^- \quad + \quad \begin{matrix} L_1 & & L_3 \\ & \overset{+}{P} & \\ L_2 & & L_4 \end{matrix} \qquad\qquad L_2^- \quad \begin{matrix} L_5 & & L_3 \\ & \overset{+}{P} & \\ L_1 & & L_4 \end{matrix}$$

ing edges are eliminated from the graph in Fig. 7 leaving the 18-vertex graph shown in Fig. 8, which resembles the carbon skeleton of hexaasterane[57]

Fig. 8

(idealized D_{6h} symmetry), and for convenience has been referred to by that term [56]. The hexaasterane graph has five different sets of edges, each set being comprised of edges which represent pseudorotation about the same pivot ligand. Each set of edges divides the 18 vertices into two subsets of nine and may be thought of as giving rise to a surface, σ_n, where n specifies the index of the edges which are associated with the surface and are bisected by it. From these considerations and from Eq. (11) one anticipates that σ_1 and σ_2 are related to intramolecular displacement reactions and that σ_3, σ_4, and σ_5 are related to intermolecular displacement reactions. Such relationships do exist and may be visualized by reference to Fig. 8. Each of the three diastereomeric pairs of enantiomers depicted in Fig. 8 straddles a σ_n and is operationally associated with it in the sense that each of the six phosphonium ions may give rise to the nine phosphoranes on its side of σ_n by intermolecular nucleophilic attack of the appropriate fifth ligand, L_n (n = 3,4, or 5). For example, phosphoranes 13, $\overline{14}$, $\overline{15}$, $\overline{23}$, 24, 25, $\overline{34}$, $\overline{35}$, and 45 comprise the western (i.e., west of σ_5) subset of diastereomeric phosphoranes and are the initial products of attack by nucleophile L_5 on the phosphonium ion ((S)– $\overset{+}{P}(L_1 L_2 L_3 L_4)$) shown on the top left of Fig. 8. With the assumption that the star-point vertices 34, $\overline{34}$, 35, $\overline{35}$, 45, and $\overline{45}$ are inaccessible and represent a virtual barrier (see Sect. 3.2), σ_1 and σ_2 fuse into a single horizontal surface, $\sigma_{1,2}$, which divides the graph into top and bottom hexagons whose vertices represent the two enantiomeric and non-interconverting sets of diastereomeric phosphoranes derived from attack of L_1 and L_2 on the two enantiomers of $\overset{+}{P}(L_2 L_3 L_4 L_5)$ and $\overset{+}{P}(L_1 L_3 L_4 L_5)$, respectively. By microscopic reversibility, each member of a particular subset of nine phosphoranes reverts to the identical phosphonium ion upon P-L_n cleavage.

Fig. 8 thus provides a useful topological representation of the stereochemical relationships of the displacement reactions in Eq. (11). In applications to chemical systems, the phosphonium ion may be chiral or prochiral [j] (as in 18, $R_1 \neq R_2$; see Sect. 3.2); however, the vertex-isomer relationship among the derived phosphoranes is dependent on which type of center is involved. If chiral, all vertices represent chiral molecules, enantiomers are related through the center of symmetry of Fig. 8 (e.g., 15 and $\overline{15}$ are enantiomers), and the phosphoranes in each subset are enantiomers of those in the partner subset

j) The term prochiral is used here (as it was before [42, 56]) with specific reference to a system such as 18 with $R_1 \neq R_2$, which is an achiral assembly containing two prochiral atoms (C-3 and P). If the prochiral carbon atom in 18 were removed ($R_1 = R_2$), then (under achiral conditions) the enantiotopic ring branches in 18 and in phosphoranes derived from it would be operationally indistinguishable; this degeneracy simplifies the hexaasterane graph into a triasterane [56, 57] of idealized D_{3h} symmetry. Enantiomers in the three dl pairs are related by a plane of symmetry perpendicular to the threefold axis, and the three achiral (C_s) conformers occupy the vertices at the points of the star which are located on that plane.

related by $\sigma_n{}^{k)}$. If prochiral[j], the vertices on the top hexagon of Fig. 8 represent six chiral diastereomers which are related to their mirror images, arranged at the vertices on the bottom hexagon, by reflection through a plane containing the points of the star (*e.g.*, $\overline{15}$ and 25 are enantiomers), which represent *meso* forms, while the phosphoranes in each subset are diastereomers of those in the partner subset related by σ_n.

3.2. Effect of Ring Constraint on the Stereochemistry of Displacement Reactions by Intermolecular Nucleophilic Attack

Nucleophilic displacement reactions which take place by attack on phosphorus in phosphonium ions have different stereochemical consequences depending on whether or not the phosphorus atom is incorporated in a small ring system. In acyclic systems, displacements usually result in inversion, whereas in small cyclic systems, retention [1] of configuration at phosphorus is usually observed. This effect was first reported by Cremer and Chorvat[59], who found that reduction of substituted phosphetane 1-oxides (*17B*) with $HSiCl_3$/triethylamine affords the corresponding phosphetanes (*17A*) with retention of configuration, in contrast to the inversion observed in the same reduction of acyclic phosphine oxides (see Sect. 2.4). Similarly, deoxygenation of *17B* with Si_2Cl_6 and base-catalyzed hydrolysis of *18B* proceed with overall retention of configuration at phosphorus (Eq. (12))[60], whereas inversion obtains for these same reactions in related acyclic systems (Eq. (5) and (7)). To assess the factors responsible for these contrasting results, a detailed analysis of the stereochemistry at the intermediate phosphorane stage was undertaken.

---

k) It should be noted that if the phosphorane contains a second chiral center, which is not "racemized" by pseudorotation, two non-interconverting enantiomeric sets exist, each containing 18 diastereomeric phosphoranes. Corresponding members of each set differ only in chirality at the second chiral center. Since enantiomers are indistinguishable under achiral conditions, either set may be chosen for analysis with the hexaasterane graph. Such an analysis has been reported [56] in connection with the base-catalyzed hydrolysis of *cis*- and *trans*-1-benzyl-1,3-dimethylphospholanium bromides (*26*, Sect. 3.2).

1) For an example of nucleophilic displacement in the phosphetane ring system which has been unambiguously determined to proceed with retention of configuration at phosphorus, see Ref. [58].

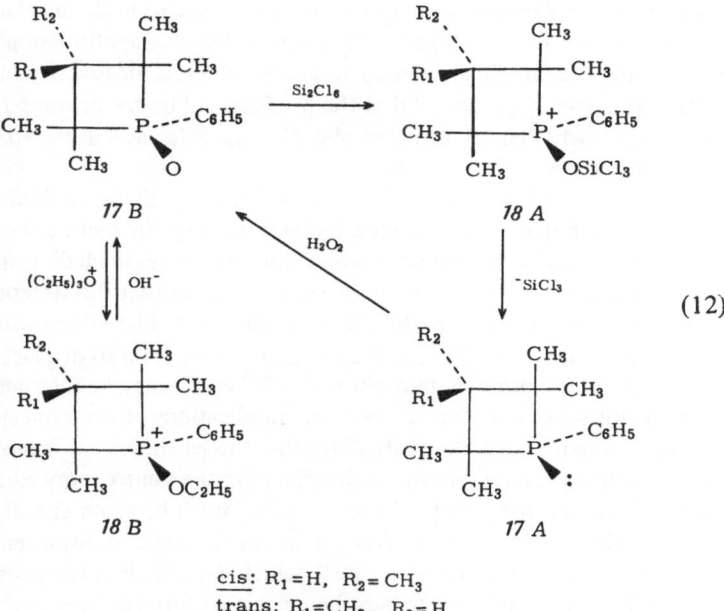

$$cis: R_1=H, \quad R_2=CH_3$$
$$trans: R_1=CH_3, \quad R_2=H$$

Phosphetanium ions *18* may be coded as shown, where L_1 and L_2 are enantiotopic ring branches, $L_3 = C_6H_5$, and $L_4 = OSiCl_3$ *(A)* or OC_2H_5 *(B)*. Random attack by L_5 ($^-SiCl_3$ or ^-OH) on *cis-* and *trans-18* yields the subsets of phosphoranes in the sectors west and east of σ_s, respectively, however a number of simplyfying assumptions may now be made. First, face *(a)* attack is preferred over edge *(e)* attack [56]; this eliminates all but $\overline{15}$, 25, $\overline{35}$, and 45 from *cis-18,* and 15, $\overline{25}$, 35, and $\overline{45}$ from *trans-18*. Second, ring strain produced when a four-membered ring is required to span the *ee* positions eliminates 35, $\overline{35}$, 45, and $\overline{45}$. Consequently, enantiomers $\overline{15}$ and 25, which are derived from attack on the enantiotopic faces of *cis-18*, and $\overline{25}$ and 15, enantiomers similarly derived from *trans-18,* are sole candidates for the initially produced phosphoranes.

cis- *18*   trans- *18*

To determine whether retention or inversion obtains in the displacement of L_4 by L_5, one need only know the sector in which the ultimate phosphorane, *i.e.*, the isomer which loses L_4 to give product, is located. By inspection of Fig. 8 it is readily perceived that the product will be *cis-* or *trans-17* depending upon whether the ultimate phosphorane is located in the southwest or northeast sector defined by σ_4.

In the process of evaluating what stereoisomers are likely possibilities for ultimate phosphoranes, the isomer number in each of the sectors related to σ_4 is reduced from nine to four by application of an "extended" principle of microscopic reversibility [61], which states in effect that the stereochemistry (*a* vs. *e*) of entry and departure must be the same. The principle of microscopic reversibility (PMR) has been extensively applied to displacement reactions at tetracoordinate phosphorus [56, 61] and it may be instructive to digress at this point in order to clarify the implications of this concept.

In mechanistic terms, the PMR states that the pathways for forward and reverse reactions at equilibrium are described by the same energy surface; it does *not* state that the profile of such a surface must be symmetrical with respect to the reaction path [62]. Application of the PMR to displacement reactions at phosphorus is aided by Fig. 9, which depicts all of the possible reaction pathways for degenerate ligand exchange at tetracoordinate phosphorus that proceed either *via* pentacoordinate transition states or *via* phosphorane intermediates capable of pseudorotation. The letters a and e in Fig. 9

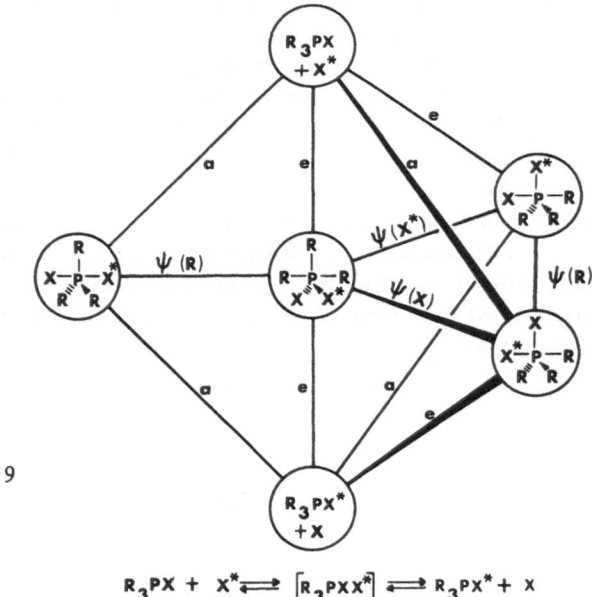

Fig. 9

$$R_3PX + X^* \rightleftarrows \left[R_3PXX^*\right] \rightleftarrows R_3PX^* + X$$

designate apical and equatorial displacement (attack or departure), respectively, of the exchanging ligand (X or X*), and the four phosphorane intermediates are interconverted by pseudorotation (ψ) about the appropriate e ligand, which is specified in the parentheses. From inspection of Fig. 9 it is readily perceived that certain reaction pathways for exchange, which may or may not include pseudorotation, have mirror symmetry while other pathways do not. However, for every unsymmetrical forward reaction pathway there exists an energetically indistinguishable reverse reaction pathway. Thus, it would be incorrect to state, without qualifications, that a attack at phosphorus by X*, followed by e departure of X, violates the PMR. This restriction applies only if the energy profile has mirror symmetry and if e attack and departure are excluded. The most that can be said from the PMR is that, assuming that bond-making and -breaking processes are rate determining, e departure is rendered unfavorable to the same extent (for symmetry reasons [62]) as a attack is preferred [56] over e attack. Moreover, if the pseudorotational processes are rate determining (for an example of such a case, see Ref. [63]), a and e departure (or *vice versa*) can no longer be excluded as mechanistic alternatives. Since a symmetric energy profile is only possible for a degenerate ligand exchange reaction, the "extended" principle (and any argument based on it) is weakened in proportion to the extent by which the attacking and leaving groups differ in character (nucleophilicity, electronegativity, etc.). Nevertheless, as a simplifying postulate, a attack and a departure will be assumed throughout this review, unless noted otherwise.

Returning to the evaluation of stereoisomers for candidates as ultimate phosphoranes, the isomer number is further reduced from four to two in each sector since ring strain effectively prevents access to the star-points (*ee*-ring). Accordingly, the ultimate phosphoranes derived from *cis-18 via* $\overline{15}$ and 25 are identified as 24 and $\overline{14}$, respectively, for retention and 14 and $\overline{24}$, respectively, for inversion. The ultimate phosphoranes are the same, but the stereochemistry of displacement is reversed, when one starts from *trans-18 via* $\overline{25}$ and 15. Since enantiomers are indistinguishable under achiral conditions, further discussion need only consider one of the two enantiomeric pathways, *e.g.,* the pathway on the top of the hexagon.

Starting from $\overline{15}$, two retention pathways exist: clockwise ($\overline{15} \rightarrow \overline{23} \rightarrow 14 \rightarrow \overline{25} \rightarrow \overline{13} \rightarrow 24$), and counterclockwise ($\overline{15} \rightarrow 24$). If it is assumed that, in the reaction under discussion, the rate of loss of L_4 is fast compared to the rates of pseudorotation, the clockwise mechanism becomes the pathway for inversion. In summary, the initial phosphorane $\overline{15}$ faces three alternatives. The first pathway ($\overline{15} \rightarrow \overline{34}$) is blocked by ring strain, and the second ($\overline{15} \rightarrow \overline{23}$) by the unfavorable [m] placement of both relatively electronegative ligands in

---

m) The generalization that electronegative substituents prefer a positions and electropositive substituents e positions may be derived empirically [48], by extension of

the e position ("stereoelectronic strain" [42]) leaving $\overline{15} \to 24$ as the only viable alternative, since this interconversion merely exchanges the a and e positions of the two relatively electronegative ligands. It follows that retention (cis-$18 \to \overline{15} \to 24 \to cis$-$17$) is the preferred pathway.

Nonempirical LCAO-MO-SCF calculations on a heuristic model system[42] have led to the same conclusions as the arguments summarized above. In addition, these same calculations revealed previously unsuspected features of the reaction system, namely, the conditions under which the star-point barrier (45) is surmountable, since it appeared[42] that the relief of stereoelectronic strain afforded when *both* relatively electronegative ligands occupy a positions (as in 45) all but compensates for ring strain. The following experimental results are in agreement with this prediction.

The relative importance of substituent electronic effects and ring strain in controlling stereochemistry was demonstrated by the comparison of the stereochemistry of the alkaline hydrolysis of *19* with that of *20*, and with that of the Si_2Cl_6 reduction of *21*. When the displaced group is a poor leav-

| 19 | 20 | 21 |

ing group and not strongly electronegative, *e.g.*, benzyl, stereochemical crossover occurs at the six-membered ring stage: alkaline hydrolysis of *19* proceeds with retention[67] while alkaline hydrolysis of *20* proceeds with partial inversion of configuration at phosphorus[68]. However, when the displaced group is a good leaving group, and with an electronegativity comparable to that of the nucleophile, the crossover point is already reached at the five-membered-ring stage: the Si_2Cl_6 reduction of *21*, a reaction analogous to the transformation in Eq. (5), proceeds with predominant inversion[67]. Thus, when the leaving group is benzyl, ring strain controls stereochemistry. However, when the attacking nucleophile (Cl_3Si^-) and the displaced group (Cl_3SiO^-) are both significantly more electronegative than alkyl or aryl, the lowest energy pathway leads, by way of a attack, to an intermediate phosphorane in which entering and leaving groups occupy a positions (45 or $\overline{45}$); relief of stereoelectronic strain more than compensates for the concomitant

---

ad hoc valence bond arguments[64], by EHMO calculations[65], or by LCAO-MO-SCF calculations on model molecules[66].

ring strain [n]. Departure of the leaving group is faster than pseudorotation and inversion obtains. For these reasons, it has been suggested [42] that the Si_2Cl_6 reduction of six-membered-ring phosphorinane 1-oxides should proceed with predominantly inversion of configuration, as in the acyclic analogues.

Factors analogous to those just discussed have been reported to explain the striking differences in stereochemistry of the alkaline hydrolysis of *22* and *23*: the former is hydrolyzed with predominant retention [70], [o] and the latter with nearly complete inversion of configuration at phosphorus [72]. In *22*, the steric effect of the *t*-butyl group replaces ring constraint as the key factor controlling the conformation of intermediate phosphoranes, and hence the overall stereochemistry of displacement [72]. Similar steric-bulk effects have been discussed with regard to the stereochemistry and product distribution of the alkaline hydrolysis of *24* [73].

$$
(CH_3)_3C \overset{\displaystyle CH_2C_6H_5}{\underset{\displaystyle C_6H_5}{\overset{|}{\underset{|}{—\ P^+ —CH_3}}}}
\qquad
(CH_3)_3C \overset{\displaystyle OC_2H_5}{\underset{\displaystyle C_6H_5}{\overset{|}{\underset{|}{—\ P^+ —CH_3}}}}
\qquad
MenO \overset{\displaystyle OC_2H_5}{\underset{\displaystyle C_6H_5}{\overset{|}{\underset{|}{—\ P^+ —CH_3}}}}
$$

22   *23*   *24*

---

n) In this connection, nucleophilic substitution of chlorine in *I* by both methoxide and benzylamine has been reported [69] to occur with *inversion* of configuration at phosphorus. These results have been rationalized by suggesting [69] *e*-attack and *e*-departure *via* a trigonal-bipyramidal intermediate with an *ae*-ring and the lone pair on phosphorus in the *a*-position. It should be noted, however, that in these two reactions the stereoelectronic demands associated with the relatively highly electronegative entering and leaving groups, and with the lone pair on phosphorus may overcome the demands of ring strain, so that in the transition state, or intermediate, the four-membered ring is forced to span the *ee*-position, with the entering and leaving groups in the *a*-position and the lone pair in an *e*-position.

$$
CH_3 \overset{\displaystyle CH_3}{\underset{\displaystyle CH_3}{\underset{\displaystyle \underset{\displaystyle Cl}{|}}{\overset{|}{\overset{\displaystyle |—CH_3}{\underset{\displaystyle —P—:}{}}}}}
$$

I

o) Base-catalyzed hydrolysis of *22* with $Na^{18}OH$, under the reported [70] reaction conditions, has been found [71] to give *t*-butylmethylphenylphosphine oxide with complete ^{18}O incorporation. This proves that attack of hydroxide ion occurs at phosphorus and excludes the possibility that attack of hydroxide ion occurs at the benzylic carbon, with displacement of *t*-butylmethylphenylphosphine which might then suffer oxidation to form the observed product.

Examples of other nucleophilic displacement reactions at tetracoordinate phosphorus, which proceed with retention of configuration and which involve systems possessing the necessary steric and electronic criteria for stereochemical analysis along the lines discussed above, include the alcoholysis [58] and aminolysis [74] of four-membered ring phosphinates and phosphinyl chlorides, respectively, and the $HSiCl_3$/triethylamine reduction of *17B* [59],p). Similar analysis of the base-catalyzed hydrolysis of *25* [75] and *26* [76] which, in contrast to the hydrolysis of acyclic analogs [77], proceed with retention

25   *26*

of configuration, is admissible [56, 73]. However, the difference between benzyl and hydroxy groups is substantial, and the possibility of direct *a* attack and *e* departure becomes a reasonable alternative in these cases.

3.3. Chemically Induced Stereomutation of Cyclic Tetracoordinate Phosphorus Compounds

Examples of chemically induced stereomutation of small phosphorus heterocycles include hydrogen chloride catalyzed epimerization of 9-phenyl-9-phosphabicyclo [4.2.1.] nonatriene [78], lithium aluminum hydride catalyzed epimerization of *17B* [79], q) and the base-catalyzed epimerization of *27* [81]. In all three examples, stereomutation may be rationalized by conversion of each substrate into a pentacoordinate intermediate *via* attack of X (Eq. (8); formally: Cl^-, H^-, and OH^-, respectively) upon a tetracoordinate phosphonium ion, followed by three pseudorotations, loss of X, and reversal to the

---

p) The original observations [59] are rationalized by extension of the analysis presented for *18A*, granted that perchloropolysilanes are the active reducing agents [34].

q) For the related lithium aluminum hydride catalyzed stereomutation of diastereomeric secondary phosphine oxides, see Ref. [80].

$$CH_3 \quad \begin{array}{c} CH_3 \\ | \\ -CH_3 \end{array}$$

$$CH_3-\!\!\!\!\!-\;\overset{+}{P}\!\!-\!CH_2C_6H_5$$

$$CH_3 \quad C_6H_5$$

27

starting substrate [r]. This sequence can be readily visualized by reference to Fig. 8.

For instance, if $X=L_5$, a attack on the phosphonium ion west of σ_5 yields 45, $\overline{15}$, 25, or $\overline{35}$ as the only possible initial phosphoranes; three consecutive pseudorotations convert these to the four possible ultimate phosphoranes 35, $\overline{25}$, 15, or $\overline{45}$, respectively, which regenerate starting phosphonium ion, with inverted configuration at phosphorus, upon loss of L_5. In each case the second pseudorotation, about L_5, crosses σ_5, which divides the stereoisomeric subsets. Note that stereomutation by this addition-pseudorotation-elimination mechanism does not require inversion of configuration of any of the intermediate phosphoranes. The only prerequisites are the maintenance of equilibrium conditions and the requirement that the rates for loss of the other ligands (*e.g.*, L_4) are slow, relative to those for the three

---

r) It has been reported that in D_2O, hydrolysis of *27*, which is slower [81] than epimerization, does not lead to deuterium incorporation at the C-3 position [75], thus ruling out epimerization mechanisms which involve stereomutation at this carbon atom. However, deuterium exchange at the benzylic carbon of *27* does occur in the presence of NaOD-D_2O and an alternative epimerization mechanism involving an ylide intermediate has been discussed [81]. Indeed, stereomutation of authentic ylide derived from *27* has been reported [75], and it has been suggested [75] that epimerization of *27* probably involves ylides and that interconversion of isomeric ylides may involve a "square planar sp^2d-hybridized phosphorus". Available data do not warrant further speculation regarding this unusual stereomutation mechanism. However, the base-catalyzed stereomutation of a phosphetanium ion incapable of ylide formation (*e.g.*, *II*) would be of considerable interest.

$$CH_3 \quad \begin{array}{c} CH_3 \\ | \\ -CH_3 \end{array}$$

$$CH_3-\!\!\!\!\!-\;\overset{+}{P}\!\!-\!C_6H_4X \qquad \text{II}$$

$$CH_3 \quad C_6H_5$$

pseudorotation steps and for the addition-elimination of X (L_5). In contra-distinction, the enantiomerization of a phosphorane whose five ligands are different requires five pseudorotations [82], and in the present cases, which involve phosphoranes incorporated in small-ring systems, this enantiomeri-zation must lead through a high-energy (*ee*-ring; star-point) intermediate.

The base-catalyzed hydrolyses of phosphetanium ions *27, 25,* and *28* [s] offer an interesting comparative series, which further illustrates the preceeding considerations and supports the suggestion of the intervention of steric-bulk effects in determining the conformational mobility of the intermediates form-ed during nucleophilic displacement reactions in these systems [72, 73].

$$CH_3$$

28

Whereas base-catalyzed epimerization of *27* is faster than hydrolysis [81], base-catalyzed hydrolysis of optically active *25* gives the corresponding P-phenyl phosphetane 1-oxide with retention of configuration [75]. The initial phosphoranes A or $\overline{A}$, which result from *a* attack of hydroxide on (*R*)- or (*S*)-*25*, respectively, can in principle equilibrate with five diastereomeric phos-phoranes, as shown in Fig. 10 [73] for (*R*)-*25* [t]. However, A, C, and E are rela-tively high energy intermediates since a *t*-butyl group is located in the *a* po-sition [73]. As discussed above, racemization prior to hydrolysis requires that loss of benzyl is slow compared to pseudorotation and addition-elimination

---

s) The synthesis of *28* was accomplished [71] using the sequence shown.

$$CH_3 \quad \xrightarrow[\text{2.) H}_2\text{O}_2]{\text{1.) 2 C}_6\text{H}_5\text{PHNa}} \quad CH_3 \quad P=O \quad \xrightarrow[\text{2.) C}_6\text{H}_5\text{CH}_2\text{Br}]{\text{1.) Si}_2\text{Cl}_6} \quad 28$$

t) In strongly alkaline media, the OH group is capable of conversion to O^-. Classifica-tion of this group as electronegative or electropositive is thus obscured; however, the conclusions based on the following argument are not affected by this problem.

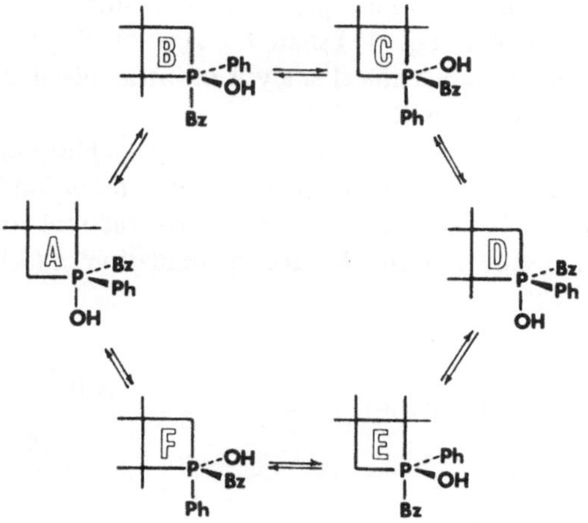

Fig. 10

of hydroxide: this condition is not met in *25* since access to the requisite ultimate phosphorane (D) is blocked by the relatively high energy intermediates C or E. Hence pseudorotation to B and loss of benzyl anion, with retention of configuration, is the preferred reaction path. In contrast, the corresponding phosphoranes derived from *27* all have *t*-butyl-like groups in the *a* position and all are therefore of comparably high energy. Thus, pseudorotation and elimination of hydroxide can successfully compete with loss of benzyl anion. Consistent with this rationale, base-catalyzed epimerization of *28* has been found[71] to be faster than hydrolysis. The corresponding phosphoranes derived from *28* do not have any *t*-butyl-like groups for placement in the *a* position and all intermediates are therefore of comparably low energy. It follows that pseudorotation and loss of hydroxide can successfully compete with P-C bond cleavage[u].

3.4. Intramolecular Displacement Reactions Proceeding Through Cyclic Intermediates

Intramolecular nucleophilic displacements at phosphorus in an acyclic system, which proceed *via* intermediate phosphoranes incorporated into a small-ring system (Eq. (11)), may result in retention of configuration. Examples of such reactions include those of benzaldehyde[77] and styrene oxide[83, 56]

---

u) The base-catalyzed hydrolysis products from *28* consist of a *ca.* 1:1 mixture of 3-methyl-1-phenylphosphetane 1-oxide and benzyl-*iso*-butylphenylphosphine oxide, the latter product resulting from ring cleavage[71].

with ylides derived from benzylphosphonium salts. Since the inter- and intra-molecular displacements in Eq. (11) share the same cyclic phosphorane intermediate, the detailed stereochemical analysis of intramolecular displacements follows the lines described in Sect. 3.2.

As an illustration, consider the oxidation of acyclic phosphines by bis (2-hydroxyethyl) disulfide [84], which proceeds with retention [56] of configuration at phosphorus. The essential portion of this reaction, and appropriate ligand indexing, are shown in Eq. (13) for the oxidation of (R)-methylphenylpropylphosphine (29).

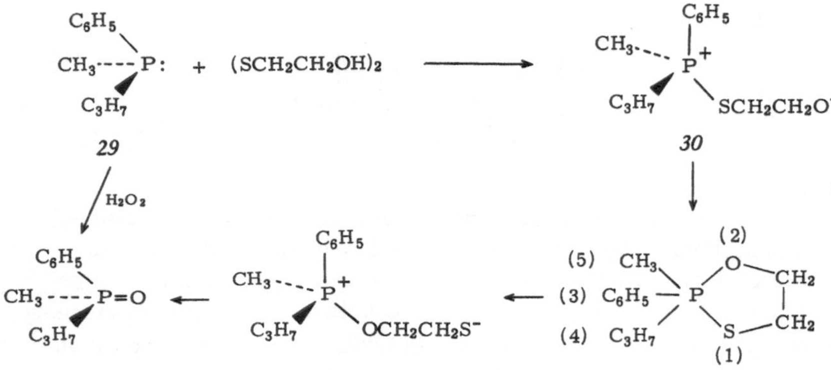

$$(13)$$

Apical attack of oxygen (L_2) on the three available faces of the intermediate phosphonium ion (30) derived from 29 leads to $\overline{23}$, 24, and $\overline{25}$, which may interconvert by pseudorotation via $\overline{13}$, 14, and $\overline{15}$, as shown in Fig. 11.

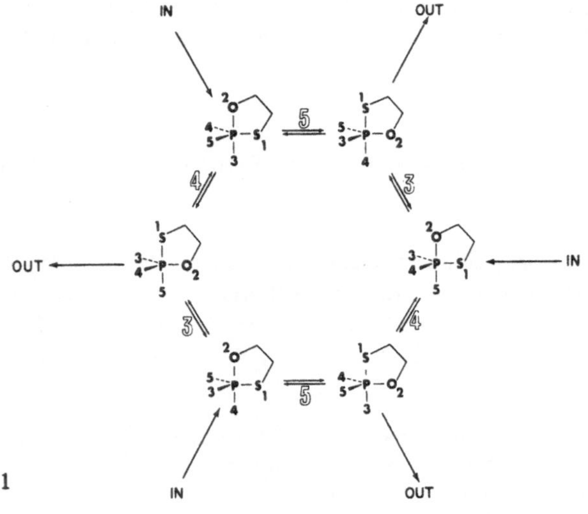

Fig. 11

Ring opening, by departure of the *a* sulfur (L_1), may occur from any one of the three ultimate phosphoranes: $\overline{13}$, 14, and $\overline{15}$. These six diastereomeric phosphoranes represented by the vertices of the top hexagon of Fig. 8, are restricted (by ring strain) from ready conversion into their enantiomers, represented by the vertices of the bottom hexagon. The overall reaction thus proceeds with retention of configuration regardless of which are the initial and ultimate phosphoranes since these six possible diastereomeric intermediate phosphoranes are all constrained to one side of $\sigma_{1,2}$, *i.e.*, they all belong to the same configurational subset.

Acknowledgement: We thank the Air Force Office of Scientific Research and The National Science Foundation for their generous support of this research.

4. References

1) Hudson, R. F., Green, M.: Angew. Chem. Intern. Ed. Engl. *2*, 11 (1963). – Horner, L.: Pure Appl. Chem. *9*, 225 (1964). – McEwen, W. E.: Topics in Phosphorus Chemistry (M. Grayson and E. J. Griffith, Ed.) *2*, Chapter 1. New York: Interscience Publishers, Inc. 1965. – Kamai, G., Usacheva, G. M.: Russ. Chem. Rev. *35*, 601 (1966). – Gallagher, M. J., Jenkins, I. D.: Topics in Stereochemistry (N. L. Allinger and E. L. Eliel, Ed.) *3*, Chapter 1. New York: John Wiley and Sons, Inc. 1968.

2) Korpiun, O., Mislow, K.: J. Am. Chem. Soc. *89*, 4784 (1967). – Korpiun, O., Lewis, R. A., Chickos, J., Mislow K.: J. Am. Chem. Soc. *90*, 4842 (1968).

3) Andersen, K. K.: Tetrahedron Letters 93 (1962).

4) Peerdeman, A. F., Holst, J. P. C., Horner, L., Winkler, H.: Tetrahedron Letters 811 (1965).

5) Lewis, R. A., Mislow, K.: J. Am. Chem. Soc. *91*, 7009 (1969).

6) Jacobus, J., Mislow, K.: Chem. Commun. 253 (1968).

7) Seyferth, D., Welch, D. E., Heeren, J. K.: J. Am. Chem. Soc. *86*, 1100 (1964).

8) Casey, J. P., Lewis, R. A., Mislow, K.: J. Am. Chem. Soc. *91*, 2789 (1969).

9) Raban, M., Mislow, K.: Topics in Stereochemistry (N. L. Allinger and E. L. Eliel, Ed.) *2*, Chapter 4. New York: John Wiley and Sons, Inc. 1968

10) Nudelman, A., Cram, D. J.: J. Am. Chem. Soc. *90*, 3869 (1968).

11) Benschop, H. P., Platenburg, D. H. J. M., Meppelder, F. H., Boter, H. L.: Chem. Commun. 33 (1970).

12) Lewis, R. A., Korpiun, O., Mislow, K.: J. Am. Chem. Soc. *89*, 4786 (1967).

13) – – – J. Am. Chem. Soc. *90*, 4847 (1968).

14) Mislow, K., Raban, M.: Topics in Stereochemistry (N. L. Allinger and E. L. Eliel, Ed.) *1*, Chapter 1. New York: John Wiley and Sons, Inc. 1967.

15) Raban, M., Mislow, K.: Tetrahedron Letters 3961 (1966).

16) Hanson, K. R.: J. Am. Chem. Soc. *88*, 2731 (1966).

17) Schulte-Elte, K. H., Ohloff, G.: Helv. Chim. Acta *50*, 153 (1967).

18) Emmick, T. L., Letsinger, R. L.: J. Am. Chem. Soc. *90*, 3459 (1968).

19) Farnham, W. B., Murray, R. K., Jr., Mislow, K.: J. Am. Chem. Soc. *92*, 5809 (1970).

20) Axelrod, M., Bickart, P., Jacobus, J., Green, M. M., Mislow, K.: J. Am. Chem. Soc. *90*, 4835 (1968). – Fleischer, E. B., Axelrod, M., Green, M., Mislow, K.: J. Am. Chem. Soc. *86*, 3395 (1964).

21) Harvey, R. G., DeSombre, E. R.: Topics in Phosphorus Chemistry (M. Grayson and E. J. Griffith, Ed.) *1*, Chapter 3. New York: Interscience Publishers, Inc. 1964.

22) Benschop, H. P., Platenburg, D. H. J. M.: Chem. Commun. 1098 (1970).

23) Farnham, W. B., Murray, R. K., Jr., Mislow, K.: Chem. Commun. 146 (1971).

24) – – – unpublished results.

25) Meppelder, F. H., Benschop, H. P., Kraay, G. W.: Chem. Commun. 431 (1970).

26) Benschop, H. P., Van Den Berg, G. R., Boter, H. L.: Rec. Trav. Chim. *87*, 387 (1968).

27) Egan, W., Tang, R., Zon, G., Mislow, K.: J. Am. Chem. Soc. *92*, 1442 (1970). – Baechler, R. D., Mislow, K.: J. Am. Chem. Soc. *92*, 4758 (1070); J. Am. Chem. Soc. *93*, 773 (1971). – Lambert, J. B., Jackson, G. F., III, Mueller, D. C.: J. Am. Chem. Soc. *90*, 6401 (1968); *92*, 3093 (1970).

[28] Horner, L., Winkler, H.: Tetrahedron Letters 461 (1964). − Baechler, R. D., Farnham, W. B., Mislow, K.: J. Am. Chem. Soc. *91*, 5686 (1969). − Baechler, R. D., Mislow, K.: J. Am. Chem. Soc. *92*, 3090 (1970).

[29] Horner, L., Winkler, H., Rapp, A., Mentrup, A., Hoffmann, H., Beck, P.: Tetrahedron Letters 161 (1961), and subsequent papers.

[30] Young, D. P., McEwen, W. E., Valez, D. C., Johnson, J. W., Vander Werf, C. A.: Tetrahedron Letters 359 (1964).

[31] Wittig, G., Cristau, H. J., Braun, H.: Angew. Chem. Intern. Ed. Engl. *6*, 700 (1967).

[32] Horner, L., Balzer, W. D.: Tetrahedron Letters 1157 (1965).

[33] Chan, T. H.: Chem. Commun. 895 (1968).

[34] Naumann, K., Zon, G., Mislow, K.: J. Am. Chem. Soc. *91*, 7012 (1969).

[35] Chan, T. H., Melnyk, A.: J. Am. Chem. Soc. *92*, 3718 (1970).

[36] Benkeser, R. A., Foley, K. M., Grutzner, J. B., Smith, W. E.: J. Am. Chem. Soc. *92*, 697 (1970). − See also Benkeser, R. A.: Accounts Chem. Res. *4*, 94 (1971).

[37] Bernstein, S. C.: J. Am. Chem. Soc. *92*, 699 (1970).

[38] Urry, G.: J. Inorg. Nucl. Chem. *26*, 409 (1964).

[39] Naumann, K., Zon, G., Mislow, K.: J. Am. Chem. Soc. *91*, 2788 (1969).

[40] Katz, T. J., Carnahan, J. C., Jr., Clarke, G. M., Acton, N.: J. Am. Chem. Soc. *92*, 734 (1970).

[41] Zon, G., DeBruin, K. E., Naumann, K., Mislow, K.: J. Am. Chem. Soc. *91*, 7023 (1969).

[42] Mislow, K.: Accounts Chem. Res. *3*, 321 (1970).

[43] Ciuffarin, E., Fava, A.: Progr. Phys. Org. Chem. *6*, 81 (1968).

[44] Haake, P., Cook, R. D., Koizumi, T., Ossip, P. S., Schwarz, W., Tyssee, D. A.: J. Am. Chem. Soc. *92*, 3828 (1970).

[45] Corfield, J. R., De'ath, N. J., Trippett, S.: Chem. Commun. 1502 (1970).

[46] Muetterties, E. L.: J. Am. Chem. Soc. *91*, 1636 (1969).

[47] Berry, R. S.: J. Chem. Phys. *32*, 933 (1960).

[48] Muetterties, E. L., Mahler, W., Schmutzler, R.: Inorg. Chem. *2*, 613 (1963). − Muetterties, E. L., Mahler, W., Packer, K. J., Schmutzler, R.: Inorg. Chem. *3*, 1298 (1964). − Muetterties, E. L., Schunn, R. A.: Quart. Rev. Chem. Soc. *20*, 245 (1966). − Schmutzler, R.: Angew. Chem. Intern. Ed. Engl. *4*, 496 (1965). − Schmutzler, R.: Halogen Chemistry (V. Gutmann, Ed.) *2*, p. 73 ff. New York: Academic Press 1967.

[49] Muetterties, E. L.: J. Am. Chem. Soc. *91*, 4115 (1969).

[50] Whitesides, G. M., Mitchell, H. L.: J. Am. Chem. Soc. *91*, 5384 (1969). − Holmes, R. R., Deiters, R. M., Golen, J. A.: Inorg. Chem. *8*, 2612 (1969).

[51] Muetterties, E. L., Storr, A. T.: J. Am. Chem. Soc. *91*, 3098 (1969). − Gielen, M., de Clercq, M., Nasielski, J.: J. Organometal. Chem. (Amsterdam) *18*, 217 (1969). − Muetterties, E. L.: Rec. Chem. Prog. *31*, 51 (1970).

[52] Coxeter, H. S. M.: Bull. Am. Math. Soc. *56*, 413 (1950).

[53] Balaban, A. T., Fǎrcasiu, D., Banica, R.: Rev. Roumaine Chim. *11*, 1205 (1966).

[54] Lauterbur, P. C., Ramirez, F.: J. Am. Chem. Soc. *90*, 6722 (1968). − Gorenstein, D., Westheimer, F. H.: J. Am. Chem. Soc. *92*, 634 (1970). − See also Cram, D. J., Day, J., Rayner, D. R., von Schriltz, D. M., Duchamp, D. J., Garwood, D. C.: J. Am. Chem. Soc. *92*, 7369 (1970).

[55] Ore, O.: Theory of Graphs, American Mathematical Society, Providence, R. I., 1962. − Harary, F.: Graph Theory, Addison-Wesley, Reading, Mass., 1969.

[56] DeBruin, K. W., Naumann, K., Zon, G., Mislow, K.: J. Am. Chem. Soc. *91*, 7031 (1969).

References

57) Biethan, U., Gizycki, U., von, Musso, H.: Tetrahedron Letters 1477 (1965).
58) Cremer, S. E., Trivedi, B. C.: J. Am. Chem. Soc. *91*, 7200 (1969).
59) — Chorvat, R. J.: J. Org. Chem. *32*, 4066 (1967).
60) DeBruin, K. E., Zon, G., Naumann, K., Mislow, K.: J. Am. Chem. Soc. *91*, 7027 (1969).
61) Westheimer, F. H.: Accounts Chem. Res. *1*, 70 (1968).
62) Burwell, R. L., Jr., Pearson, R. G.: J. Phys. Chem. *70*, 300 (1966).
63) Kluger, R., Covitz, F., Dennis, E., Williams, L. D., Westheimer, F. H.: J. Am. Chem. Soc. *91*, 6066 (1969).
64) Walsh, A. D.: Discussions Faraday Soc. *2*, 18 (1947). — Bent, H. A.: Chem. Rev. *61*, 275 (1961).
65) Van Der Voorn, P. C., Drago, R. S.: J. Am. Chem. Soc. *88*, 3255 (1966).
66) Rauk, A.: unpublished work (Princeton).
67) Egan, W., Chauvière, G., Mislow, K., Clark, R. T., Marsi, K. L.: Chem. Commun. 733 (1970).
68) Marsi, K. L., Clark, R. T.: J. Am. Chem. Soc. *92*, 3791 (1970).
69) Smith, D. J.H., Trippett, S.: Chem. Commun. 855 (1969).
70) De'ath, N. J., Trippett, S.: Chem. Commun. 172 (1969).
71) Zon, G.: unpublished results.
72) Lewis, R. A., Naumann, K., DeBruin, K. E., Mislow, K.: Chem. Commun. 1010 (1969).
73) DeBruin, K. E., Mislow, K.: J. Am. Chem. Soc. *91*, 7393 (1969).
74) Hawes, W., Trippett, S.: J. Chem. Soc. (C) 1465 (1969).
75) Corfield, J. R., Shutt, J. R., Trippett, S.: Chem. Commun. 789 (1969).
76) Marsi, K. L.: Chem. Commun. 846 (1968); J. Am. Chem. Soc. *91*, 4724 (1969).
77) Bladé-Font, A., VanderWerf, C. A., McEwen, W. E.: J. Am. Chem. Soc. *82*, 2396 (1960). — McEwen, W. E., Kumli, K. F., Bladé-Font, A., Zanger, M., VanderWerf, C. A.: J. Am. Chem. Soc. *86*, 2378 (1964).
78) Katz, T. J., Nicholson, C. R., Reilly, C. A.: J. Am. Chem. Soc. *88*, 3832 (1966).
79) Henson, P. D., Naumann, K., Mislow, K.: J. Am. Chem. Soc. *91*, 5645 (1969).
80) Farnham, W. B., Lewis, R. A., Murray, R. K., Jr., Mislow, K.: J. Am. Chem. Soc. *92*, 5808 (1970).
81) Cremer, S. E., Chorvat, R. J., Trivedi, B. C.: Chem. Commun. 769 (1969).
82) Muetterties, E. L.: Inorg. Chem. *6*, 635 (1967).
83) McEwen, W. E., Bladé-Font, A., Vander Werf, C. A.: J. Am. Chem. Soc. *84*, 677 (1962). — McEwen, W. E., Wolf, A. P., VanderWerf, C. A., Bladé-Font, A., Wolfe, J. W.: J. Am. Chem. Soc. *89*, 6685 (1967).
84) Grayson, M., Farley, C. E.: Chem. Commun. 831 (1967).

Received January 20, 1971

SPRINGER-VERLAG
BERLIN·HEIDELBERG·NEW YORK

Herausgegeben von
H. Mayer-Kaupp

Anleitungen für die chemische Laboratoriumspraxis

Band 1 : Seith/Ruthardt
Chemische
Spektralanalyse

Eine Anleitung zur Erlernung und Ausführung von Emissions-Spektralanalysen

Von Wolfgang Seith und Konrad Ruthardt
Sechste, ergänzte Auflage von Dr. Walter
Rollwagen, o. Professor für Experimentalphysik
an der Universität München

Mit 84 Abbildungen
und einer Tafel
XII, 185 Seiten. 1970
Gebunden DM 48,—
US $ 13.20

Aus den Besprechungen: „Die 5. Auflage des Werkes von Seith/Ruthardt hat von dem neuen Bearbeiter ein völlig anderes Gesicht erhalten. Es ist nicht nur der raschen apparate-technischen Entwicklung auf diesem Gebiet seit Erscheinen der 4. Auflage 1949 Rechnung getragen worden, sondern es ist bewußt die Behandlung der experimentellen Grundlagen wesentlich ausführlicher gestaltet worden. Besonders breiten Raum nehmen Hinweise auf den Umgang mit Spektralapparaten, Projektoren, Photometern und photographischen Platten ein. Es werden durchweg neuere Geräte gezeigt. Das Büchlein hat auch in dieser Form nicht wesentlich an Umfang zugenommen und wird von jedem, der mit Spektralanalysen zu tun hat, gern als Ratgeber zur Hand genommen werden." *Zeitschrift für analytische Chemie*